OF PANDAS AND PEOPLE

The Central Question of Biological Origins

Second Edition

Percival Davis
Dean H. Kenyon

Charles B. Thaxton
Academic Editor

HAUGHTON PUBLISHING COMPANY
DALLAS, TEXAS

Cover photograph: © Steve Kaufman/Peter Arnold, Inc.
Book illustration: Debbie Smith

Printed in the United States of America.

ISBN 0-914513-40-0

ACKNOWLEDGEMENTS

Of Pandas and People went through an evolution of its own. The book's Project Chairman and Academic Editor, Dr. Charles B. Thaxton, supervised the review and revision process through many drafts. Over an extended period of time, the manuscript, either in part or in its entirety, was sent to scores of reviewers with very diverse perspectives. In addition, the book was used during its development for two years in a public school district in field test form, and feedback was taken into account in further revision. Although the manuscript was nearly always under review by someone, there was a general cadence to these reviews, which came in three basic rounds. First came a round by scientists engaged in teaching and research, then a round by high school teachers, then a second round by scientists. Many hundreds of valuable criticisms and suggestions were offered, from readers holding evolutionary views as well as others in general agreement with the authors. Upon publication of the first edition, additional valuable comments became available from a wider scientifically informed readership.

Our genuine thanks to the following for their indispensable assistance in the review of portions or all of the manuscript; any errors of fact or interpretation are not theirs but our own:

Critical Reviewers

Robert L. Agnew
Dean of Natural and Social Sciences
North Lake College
Irving, Texas

Robert M. Augros
Professor of Philosophy
St. Anselm College
Manchester, New Hampshire

Donald M. Austin
Former President of Anthropology
Southern Methodist University
Dallas, Texas

Richard A. Baer, Jr.
Professor, Dept. of Natural Resources
Cornell University
Ithaca, New York

Arthur L. Battson, III
Director of Instructional Resources
University of California
Santa Barbara, California

John R. Baumgardner
Technical Staff, Theoretical Division
Los Alamos National Laboratory
Los Alamos, New Mexico

Michael J. Behe
Assoc. Professor of Chemistry
Lehigh University
Bethlehem, Pennsylvania

Raymond G. Bohlin
Director of Research
Probe Ministries, Intl.
Richardson, Texas

Walter L. Bradley
Professor and Head
Dept. of Mechanical Engineering
Texas A&M University
College Station, Texas

James O. Buswell, III
Vice President for Academic Affairs
The William Carey Intl. University
Pasadena, California

Donald F. Calbreath

Harold G. Coffin
Former Research Scientist
Geoscience Research Institute
Loma Linda University
Loma Linda, California

Turner Collins
Professor of Biology
Evangel College
Springfield, Missouri

Thomas L. Compton

Thomas C. Emmel
Professor of Zoology
University of Florida
Gainesville, Florida

Norman L. Geisler
Dean
Southern Evangelical Seminary
Charlotte, North Carolina

L. James Gibson
Research Scientist
Geoscience Research Institute
Loma Linda University
Loma Linda, California

Charles Hagar
Professor of Astronomy
San Francisco State University
San Francisco, California

Lois Harbaugh
Science Teacher/Chairperson
Lake Highlands Junior High
Dallas, Texas

Arnold Hyndman
Assoc. Professor, Biological Sciences
and Assoc. Provost
Rutgers University
New Brunswick, New Jersey

T. Rick Irvin
Assoc. Professor of Environmental Studies
Louisiana State University
Baton Rouge, Louisiana

Robert Kaita
Principal Research Physicist
Plasma Physics Laboratory
Princeton University
Princeton, New Jersey

Alexander Mebane
Research Organic Chemist

Stephen C. Meyer
Asst. Professor of Philosophy
Whitworth College
Spokane, Washington

Gordon C. Mills
Emeritus Professor of Biochemistry,
Dept. of Human Biological Chemistry and Genetics
University of Texas Medical Branch
Galveston, Texas

J. P. Moreland
Professor of Religion
Talbot School of Theology
La Mirada, California

Tim Morrill
High School Science Instructor

Donald Munro
Dept. of Biology Chair
Houghton College
Houghton, New York

Paul A. Nelson
Ph.D. candidate
University of Chicago
Chicago, Illinois

Kenneth O'Loane
Research Chemist

S. D. Palmer

Alvin Plantinga
John A. O'Brien Dept. Chair
and Professor of Philosophy
University of Notre Dame
Notre Dame, Indiana

J. David Price
Former Professor of Science Education
University of California at Los Angeles
Los Angeles, California

Harry J. Reynolds
Superintendent
Chattanooga Public Schools
Chattanooga, Tennessee

David M. Shotton
Professor of Biology
Oxford University
Oxford, England

Fred Sigworth
Professor of Physiology
Yale University
New Haven, Connecticut

Lee Spencer
Paleontologist
Bureau of Land Management
Salt Lake City, Utah

Peter Vibert
Senior Scientist, Rosenstiel Center
Brandeis University
Waltham, Massachusetts

John L. Weister
Chairman, Committee for Integrity in Science Education
American Scientific Affiliation
Buellton, California

Harold T. Wiebe
Dean of Graduate School
Professor of Biology
Seattle Pacific University
Seattle, Washington

David L. Wilcox
Chairman, Biology Department
Eastern College
St. Davids, Pennsylvania

Kurt Wise
Asst. Professor of Science
Bryan College
Dayton, Tennessee

Editors and Contributors

Nancy R. Pearcey
(Overview Chapter)
Rod Clark
A. James Melnick
Joseph E. O'Day
Audris Zidermanis
Gordon E. Peterson
David N. Quine

CONTENTS

User's Guide to the Book

The authors and editors of this book have provided several special features to make it as flexible and easy to use as possible.

Subjects on Two Levels

They have written the book, for example, on two levels. Lighter, easy to understand treatments of the subjects are found in the first chapter, "Of Pandas and People: An Overview." More in-depth treatments of the same material are provided in the Excursion chapters that follow. The authors have included the in-depth material because it is easy to sacrifice accuracy when trying to make a scientific subject easy to follow, and this is a subject where accuracy is important.

If you plan to read one or more of the Excursion chapters, be sure to read the matching Overview section or sections first. (Section 1 of the Overview is elaborated on in Excursion Chapter 1, Section 2 in Excursion Chapter 2, etc.) You will find the more challenging Excursion chapters especially helpful if you plan to do a special project or paper on some subject area covered by the book.

Vocabulary

Glossary

Even if you find the Excursion chapters challenging, there are several helps provided. You will notice that some words are in boldface. Any boldfaced word is defined for you in the Glossary at the end of the book. This same word might be used several times, but it is only in boldface the first time it is used in the Overview chapter and the first time it is used in the Excursion chapters. If you have any doubts about the meaning of a word in boldface as you are reading, check it out then.

Pronunciation

Before you have read very far, you will come upon a pronunciation guide in parentheses for a technical word that is not easily pronounced unless you have heard it before. These pronunciation guides are easy to follow, and will help you use the terms in discussion as well as to "hear" them correctly, even when reading silently.

Emphasis

You will also see certain words in italics. This may be for either of two reasons: 1) To emphasize a term, either technical or general, in order to help you get at the heart of the subject. Technical terms in italics are usually explained or defined right in the discussion. In cases where italics are used for emphasis, the word is only italicized in its first appearance in the Overview chapter, and once again, at most, in the Excursion chapters. 2) A second purpose of italics is for genus and species names of organisms. In these cases, it is used each time they are named.

Easy Access to Subjects

Index

An index is provided in the back, so that if you want to find out what the book says about a specific subject, whether you have already read about it in the book or not, you can find the pages to turn to for that subject.

Reference to Main Textbook

Occasionally the authors will refer you to your main textbook concerning a subject. If you feel unsure about that subject or concept, it might be worthwhile for you to look it up in the index of your main text and review it briefly.

Lists of References

Since the Excursion chapters surpass the Overview chapter in depth of treatment, references to scientific literature are listed at the end of each Excursion chapter and of A Note to Teachers.

Introduction

> **We live in an extraordinary age. These are times of stunning changes in social organization, economic well-being, moral and ethical precepts, philosophical and religious perspectives, and human self-knowledge, as well as in our understanding of that vast universe in which we are embedded like a grain of sand in a cosmic ocean. As long as there have been human beings, we have posed the deep and fundamental questions, which evoke wonder and stir us into at least a tentative and trembling awareness, questions on the origins of consciousness; life on our planet; the beginnings of the Earth; the formation of the Sun; the possibility of intelligent beings somewhere up there in the depths of the sky; as well as, the grandest inquiry of all—on the advent, nature and ultimate destiny of the universe.[1] —Dr. Carl Sagan**

Carl Sagan, one of the foremost popularizers of science in our time, has drawn our attention to ancient, important, and fascinating questions. How did this immense universe come into existence? How did the earth come to harbor life? What does it all mean, if anything, and how do mere mortals like ourselves fit into the overall scheme of things, if indeed there be a scheme? As Dr. Sagan reminds us, we are not the first to wonder, nor are we likely to be the last.

Two different concepts of the origins of living things have long histories extending from ancient times to the present. While both have taken varied forms through the centuries, there is, nevertheless, a central core idea that modern proponents of each view hold in common with their forebears. Through all the ages some have held the concept of life emerging from simple substance. What the substance is, what form the first life took, and the mechanism of emergence, chance or law, are details that have changed to characterize many different theories of natural origin. Likewise, proponents of intelligent design throughout history have shared the concept that life, like a manufactured object, is the result of intelligent shaping of matter. Within intelligent design also, the details as to how gradual or abrupt, and over what span of time, differ.

Despite all society has learned, answers given for the big questions remain in conflict or uncertain. Uncertainty keeps science—a quest for knowledge—alive. We should be thankful for those who have been brave enough to persist in asking important questions, for they have breathed life into our existence.

This book has a single goal: to present data from six areas of science that bear on the central question of biological origins. We don't propose to give final answers, nor to unveil *The Truth*. Our purpose, rather, is to help readers understand origins better, and to see why the data may be viewed in more than one way. There will be no attempt to kid you—to tell you that a complex issue is simple, or that the authors' view is the only reasonable one.

From these six areas of science, we will present interpretations of the data proposed by those today who hold the two alternative concepts; those with a Darwinian frame of reference, as well as those who adhere to intelligent design. We will concentrate, however, on explaining what few other textbooks do: the scientific rationale of the second concept. Our intention has been to give you presentations that will balance the biology curriculum. For what might be a refreshing change, you are asked to form your own opinions. If you understand the information presented, you are fully capable of drawing your own conclusions.

You will be given an opportunity to examine data dealing first with how life may have arisen. Once that has been addressed, you can delve into what science has to say about the impact of genetics and environment on shaping groups of organisms that we commonly refer to as species. This raises the questions of how species came into being in the first place, and how the elegantly complex structures of organisms arise. Have you ever wondered how we can explain the human eye, or the marvel of flight in birds?

Enough fossils of once-living organisms are found near the earth's surface to collect in great numbers. The fossils are known in such abundant supply that a century ago, the Smithsonian Institution gave away almost 25,000 "leftover" specimens of marine invertebrate fossils in just one year. Today, many millions are available for examination. Do they have a story to tell?

Evidences of creatures that once lived but are no longer with us abound. One has only to visit areas such as Dinosaur National Monument, or other fossil beds to be staggered by the variety and enormous size of these extinct forms, creatures that no longer roam the face of our planet, except in the woefully unscientific productions of television and movies.

When we examine the vertebrates, animals with bony skeletons, we see repeating patterns in forelimbs and hindlimbs, even though the proportions of the various bones differ considerably between organisms, and the limbs themselves have different functions. What can such patterns mean, and how are they to be explained? Most agree without hesitation that there are patterns. The problem arises when it comes to interpretations of cause: mere probability, "family resemblance," or the product of intelligent design?

The last half century has permitted science to explore more than flesh and bone, fascinating as they are. It is now possible to compare and contrast organisms, be they plant, animal, or other, by their body chemistry. Proteins, a vast army of molecules with a staggering array of roles to play in living things, reveal intriguing similarities and equally intriguing differences. Evidence to date forces us to conclude they are produced only by living things. Do these speak of random natural causes, or of something intelligent?

We live in an exciting time. We have the technology that makes possible more observations than were available to any who lived before us. Coupled with these, we have intelligence enough to propose sophisticated explanations for what we see. But let us be honest: we do not have all the answers, and we will not in the near future, despite what some would have you believe. Science can provide us with additional answers, however, if used patiently, and if understood. Science includes many elements; it includes asking what causes things.

In the world around us we observe two classes of things: natural objects, like stars and mountains, and man-made creations, such as houses and computers. To put this into the context of origins, of how things arose, we see things resulting from two fundamentally different causes: natural and intelligent.

If we didn't see it occur, how do we decide whether something resulted from natural or from intelligent causes? We do it without giving it much thought. We see clouds and we have decided that they are the result of natural forces. No matter how intricate their shapes may be, we understand that all clouds are simply water vapor affected by wind and temperature.

Walking along a beach you may be impressed by the regular pattern of ripples in the sand. The scene may be artistic but it isn't likely that you would look around for an artist who might be responsible. A natural cause, you rightly conclude. But if you come across words unmistakably reading "John loves Mary" etched into the sand, you would know that no wave action was responsible for that. Nor would you be likely to imagine that, given enough time, grains of sand would spontaneously organize themselves so uniquely. Rather, you would look around for an intelligent cause: John . . . perhaps even Mary.

What do these examples suggest? The way we decide whether a given phenomenon arises from natural or intelligent causes is from experience, refined as we mature and are exposed to more and more examples. Through this process, our experience grows into a collection of uniform observations, things we learn to count on. If experience has shown that a certain class of phenomena results from intelligent causes and then we encounter something new but similar, we conclude its origin also to be from an intelligent cause. This is what must happen, for example, if scientists are to recognize evidences of intelligent beings elsewhere in the universe, as Carl Sagan speculated they might. *Of Pandas and People* presents evidences, found in the data of biology, for intelligent cause. We hope its presentations are interesting, honest, and not overstated.

Recently the U.S. Supreme Court made it clear that teachers have the right to present non-evolutionary views in their classrooms. This is significant, since a survey of high school biology teachers in Ohio at about the same time showed that roughly one fifth were already presenting the idea of creation in their science classes, and a quarter of the teachers surveyed believed that the subject should be included in science classrooms.[2] Other surveys of science teachers, including a nationwide market study done for this book,[3] confirm the same or greater levels of interest among teachers.

The authors and publisher want you to use this book as a supplement, not a substitute, for your biology text; it cannot replace the main textbook. But without *Of Pandas and People*, you would miss a lot of interesting science. We hope you finish this book respecting good scientists of all persuasions; we do. The subjects here are treated in depth, and digging deeper brings richer rewards. Your textbook provides a lighter treatment of a broader range of topics. Wander back and forth between the two, using each to enrich the other.

This publication has a special design. The first chapter, "*Of Pandas and People*: An Overview," treats all of the major topics of the book. It gives you the option of gathering only the essence of the six areas covered, should time or interest be limited.

You may read this chapter on its own, or as an introduction to the more detailed, more academically challenging discussion of the subjects in the six Excursion chapters that follow.

Perhaps you have already discovered the rewards of learning, and you want to "go for it"—to dig deeper so you will understand more. In that case you will want to read all of "An Overview" and each of the six Excursion chapters. These chapters have references to help you go even deeper if you want to.

Of Pandas and People is not intended to be a balanced treatment by itself. We have given a favorable case for intelligent design and raised reasonable doubt about natural descent. But used together with your other text, it should help to balance the overall curriculum. By now you are aware that you have a mind of your own. Here is a good opportunity to use it. We hope that you will enjoy dealing with ideas and coming to conclusions on your own. Of course, you may modify your understandings as you learn more in the years to come. Meanwhile, we expect that *Of Pandas and People* could become an exciting event in your educational journey.

---

1. Carl Sagan, 1974. *Broca's Brain*. New York: Random House, from the Introduction.
2. Michael Zimmerman, 1987. "The Evolution-Creation Controversy: Opinions of Ohio High School Biology Teachers," *Ohio Journal of Science*, 1987 4: pp. 115-125.
3. Available from *Foundation for Thought and Ethics*, P.O. Box 830721, Richardson, Texas 75083-0721.

Of Pandas and People: An Overview

OVERVIEW SECTION 1

The Origin of Life

There was a time when people believed some animals arose on their own, full-blown, from non-living matter. The belief was called "**spontaneous generation**." Today the idea might seem to be no more than superstition, but at one time it seemed to be confirmed by common-sense experience. Leave rotting meat out, and isn't it quickly covered with maggots? Leave dirty rags in the corner of a shed, and doesn't it soon become a nest of mice?

With the rise of scientific method, however, belief in spontaneous generation began to be discredited. In 1668 Francesco Redi conducted an experiment to determine whether worms arose spontaneously in decaying food. He placed similar samples of raw meat in two sets of jars. One set he covered with a muslin screen, the other he left open. After several days, the muslin screen covering the first sample was sprinkled with fly eggs, but there were none on the meat itself. The meat in the open jar was covered with eggs, which soon hatched into maggots. Redi had shown that maggots were not simply small worms that arose spontaneously, but rather were fly larvae. He had demonstrated scientifically that at least some life can come only from life.

As the idea of spontaneous generation began to wane, its last outpost was the world of microscopic life. The microscope had revealed the existence of a world hitherto invisible and unsuspected. Microscopic creatures were so small, and appeared to be so simple, that it was not difficult to believe they arose spontaneously from nonliving matter. After all, if bits of straw were left to rot in a pan of water, the water was soon swarming with bacteria.

In the early 1860s, Louis Pasteur laid this notion to rest as well. He showed that water could be kept free of bacteria by boiling it and then exposing it to only purified air. By doing so, he proved that the microscopic life that mysteriously appeared as the straw rotted was the result of airborne bacteria.

At the same time Pasteur was doing his research, Charles Darwin and others were formulating the mechanistic theory of evolution. Pasteur's discoveries

Louis Pasteur, who proved that the ideas of spontaneous generation of his time were not valid.

were all very well for the advance of scientific knowledge, but they were somewhat disturbing for the notion of a purely natural origin. Redi and Pasteur had shown that full-blown organisms do not arise from nonliving materials, whether mice or microbes. Yet Darwinism, the dominant theory of evolution, seemed to require some type of natural origin. How was this possible?

Then, in the 1920s, a Russian scientist named A.I. Oparin suggested a new approach to the question. Life did originate from nonliving matter, he proposed, but not all at once. Instead, it arose very gradually in a series of stages. As biochemists have studied the chemistry of life, details of these stages have since also been suggested: simple chemicals combined to form organic compounds, such as amino acids, which in turn, combined to form large, complex molecules, such as proteins, which aggregated to form an interconnecting network and a cell wall. Oparin's hypothesis is referred to as **prebiotic evolution** (prebiotic means "before life") or **chemical evolution** (because it assumes life began in a sea of chemicals called the *prebiotic soup*). The Oparin hypothesis has become the standard evolutionary approach to the origin of life. Oparin and subsequent scientists suggested that chance interactions of chemicals and compounds would not eventually lead to viable molecules. Scientists no longer feel that chance is enough. Instead, they theorize that some internal tendency of matter gives rise to the ordered structure we see in living compounds. According to today's view, matter has within itself a tendency toward self-organization leading to life.

Life in a Test Tube

One of the advantages of Oparin's hypothesis was that, to some extent, it could be tested. Not directly tested, of course, because we cannot subject a past event like the origin of life to observation. Instead, scientists can construct stories about the events that may have happened, and then set up laboratory experiments to see if similar events will occur today. These are called simulation experiments, because they are designed to simulate what might have happened on the early earth when life began.

Oparin suggested that life arose from chemical reactions among simple gases in the atmosphere: methane, ethane, ammonia, hydrogen, and water vapor. These reactions were activated by various forms of energy in the environment—by lightning, heat from volcanos, kinetic energy from earthquakes, or light from the sun. When they encountered this energy, the atmospheric gases were converted into more complicated compounds to form amino acids, fatty acids, and sugars. These accumulated in the primitive oceans until there were enough to link up and form even larger, more complex molecules, such as proteins and

DNA. Eventually these molecules combined to form integrated particles called **coacervates** (co-AS-er-vates, liquid drops or bubble-like structures). From these there arose the first true cell, complete with cell membrane, complex metabolism, genetic coding, and the ability to reproduce.

How has this hypothesis been tested in the laboratory? Scientists have taken the simple gases suggested by Oparin, mixed them together and subjected them to various energy sources, such as ultraviolet light (to simulate sunlight) and electrical discharges (to simulate lightning). In 1953, Stanley Miller and Harold Urey reported the first such experiment. A gooey, tar-like substance formed in the flask. Examining it, Miller identified several of the amino acids found in proteins today. Since then, other biological compounds have been detected in similar atmospheric simulation experiments. The list now includes most of the essential kinds of organic compounds found in living things.

Stanley Miller, shown here with the apparatus used in his original origin of life experiments as a young graduate student.

Unnatural Conditions

Imagine the excitement in the scientific community as the results of these experiments were first published. The success of such early experiments greatly increased the credibility of evolutionary theory. But when scientists sought to go beyond the simplest building blocks of life, the momentum slowed. The step from simple compounds to the complex molecules of life, such as protein and DNA, has proved to be a difficult one. Thus far, it has resisted all efforts by the scientists working on the problem.

The problem is that some chemical reactions occur quite readily, whereas others do not. The simple building blocks of life form relatively easily. They form in reactions belonging to the classes that occur readily. But the chemical reactions required to form proteins and DNA do not occur readily. In fact, these products haven't appeared in any simulation experiment to date.

In addition, some scientists have been very critical of some of the assumptions made in constructing simulation experiments. The procedures of such experiments should, after all, simulate what might reasonably have happened under natural conditions. Yet many don't.

Oxygen in the Air

For example, all experiments simulating the atmosphere of the early earth have eliminated molecular oxygen. The reason is that oxygen acts as a poison preventing the chemical reactions that produce organic compounds. Furthermore, if any chemical compounds did form, they would be quickly destroyed by oxygen reacting with them, a process called oxidation. (Many food preservatives are simply substances that protect food from the effects of oxidation.)

For these reasons, the standard story of chemical evolution assumes that there was no oxygen present in the earth's atmosphere at the origin of life. Yet scien-

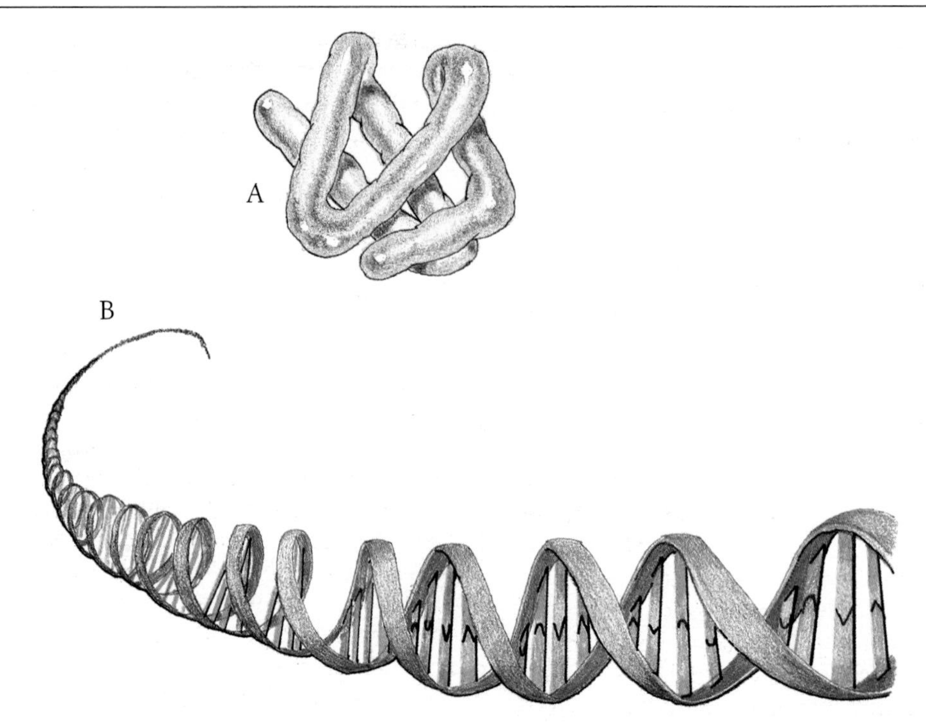

Protein and DNA molecules. The protein molecule (A) is shown in its folded up form, which is secured by chemical and physical forces between its parts, and therefore different for each individual protein. The DNA (B) is shown in its famous double helix (corkscrew) form.

tists now have strong evidence that oxygen was present on the earth from the earliest ages. Many minerals react with oxygen (like the rusting of iron), and the resulting oxides are found in rocks dated earlier than the development of life according to the hypothesis. Moreover, significant levels of oxygen would have been necessary to produce ozone which would shield the earth from levels of ultraviolet radiation lethal to biological life. Clearly, a realistic simulation of the early earth's atmosphere would have to include oxygen.

Reversible Reactions

A second barrier faced by any theory of chemical evolution can be stated as a paradox. Some chemicals react quite readily with one another. They connect easily, like the north and south poles of two magnets. Others resist reacting. Getting them to react is like forcing the magnets' north ends together. To drive such a chemical reaction forward requires energy (heat, for example, or electricity). But—and here is the paradox—energy also breaks chemical compounds apart. In fact, energy is actually more likely to break them apart than to cause them to bond in the first place.

When a chemist exposes a mix of chemicals to heat or electricity, some compounds may form but others will break down. Since the process of destruction is actually more likely to occur, the net result will be only a small amount of chemical compounds. Those that do form will generally be simple ones, since any complex molecules that might form would quickly break back down to their simpler components.

In simulation experiments, amino acids and other products that form are siphoned off through a trap in order to protect them from breaking back down again. But there is no evidence of anything in nature that acts as a counterpart to such a trap. Results reported from experiments with traps are therefore overblown. The procedures of such experiments should, after all, duplicate what might reasonably have happened under natural conditions. Yet many don't. Those that don't undermine the whole purpose of trying to simulate nature.

Chemistry in 3-D

Amino acids, sugar, protein, and DNA are not simply strings of chemicals. For one thing they exhibit very specific three-dimensional structures. When synthesized in the laboratory, they may have the right chemical constituents but still exhibit the wrong three-dimensional form.

For example, amino acids appear in two forms that are mirror images of each other (see Figure 1) just as a right glove is the mirror image of a left glove. The two forms are referred to as right- and left-handed amino acids. Living things use only left-handed amino acids in their proteins. Right-handed ones don't "fit" the metabolism of the cell any more than a right-handed glove would fit onto your left hand. If just one right-handed amino acid finds its way into a protein, the protein's ability to function is reduced, often completely. (As it happened, only right-handed sugars are used in DNA and RNA.)

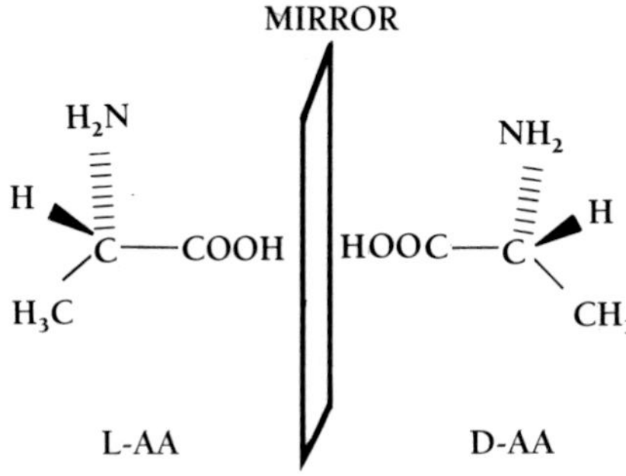

Researchers have found no natural conditions they can incorporate in simulation experiments that will produce only the correct three-dimensional structure. When amino acids are synthesized in the laboratory, the result is an equal mixture of both forms, like a pile of right- and left-handed gloves. In this and other ways, life shows characteristics that appear to be alien to anything produced under natural conditions.

Cross-Reactions

The fact that some reactions occur readily whereas others do not creates another problem. As we have said, the reactions involved in the formation of biologically important compounds are the kind that have to be made under artificial conditions. Amino acids, for example, do not readily react with each other. A corollary is that they do react readily with other substances. Herein lies a problem. If amino acids formed on the early earth, they would not float around in lakes and ponds simply waiting for the right partner amino acids to show up in order to form proteins. Instead, they would combine with other compounds in all sorts of cross-reactions, tied up and unavailable for any biologically useful function (see Figure 2). This explains why, in actual experiments, the predominant outcome is large yields of non-biological goo.

To summarize, most simulation experiments do not actually mimic in a realistic way the conditions that would have existed on the early earth. Taking into account the presence of oxygen, the fact of reversible reactions, the prevalence of cross-reactions, and the lack of uniform three-dimensional structures, we must conclude that conditions on the early earth were far from favorable for the spontaneous emergence of living things. The most probable scenario of early earth history is not one of synthesis but of destruction. The experiments do not support the idea that there exists an inherent tendency toward self-organization within the starting materials.

MIRROR

H_2N / H / C—COOH / H_3C | HOOC—C / NH_2 / H / CH_3

L-AA | D-AA

Figure 1. *Amino acids come in two forms "right-handed" (D-) and "left-handed"(L-). Both forms of the amino acids alanine are shown here.*

The Language of Life

We often get the impression from newspaper articles and even some textbooks that scientists have come close to "creating life" in the laboratory. Yet compounds synthesized in the laboratory fail to exhibit the special sequences or the three-dimensional structure necessary for biological functioning.

Though life is made of simple chemicals, we should not conclude that living things are themselves simple. Shakespeare's sonnets are artistically complex, though they are composed of simple letters made to form words and phrases. Mozart's pieces are musically complex, though they consist of simple notes placed in patterns. The decisive factor in living things is not the simple components but the patterns.

What patterns are essential to life? Since the discovery of the DNA molecule, most scientists define life with reference to the coded information in the DNA. It is this information that governs the development and functioning of all the cells in our bodies.

The origin of life, then, includes the origin of coded information. The large molecules crucial to life, such as protein and DNA, are constructed much like a message in a known language, with chemicals acting as letters and combining in defined sequences to form words, phrases, and sentences. The "message" is decoded by the cell much the same way the dots and dashes of messages in Morse Code can be decoded by anyone who knows it.

In the world around us, we see two classes of things: natural objects, like rivers and mountains, and man-made structures, like houses and computers. To put it in the context of origins, we see things resulting from two kinds of causes: natural and intelligent.

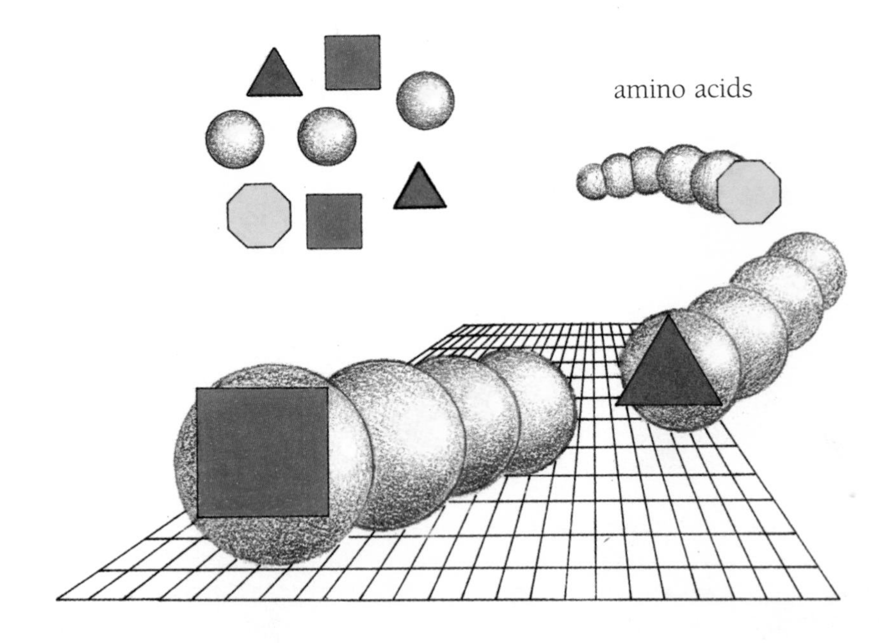

Figure 2. ***Cross reactions. In a primordial soup, nothing would prevent chains of amino acids from reacting with innumerable other ingredients of the primordial soup, thus tying them up and ending their potential to become useful protein molecules.***

Uniform Experience

How do we decide whether something is the result of natural processes or **intelligent causes**? Most of us do it without even thinking. We see clouds and we know, based on our experience, they are the result of natural causes. No matter how intricate the shapes may be, we know that a cloud is simply water vapor shaped by the wind and the temperature. On the other hand, we may see something looking very much like a cloud that spells out the words "Vote for Smedley." We know that, even though they are white and fluffy like clouds, the words cannot be the result of natural causes.

Why not? Because our experience—and that of everybody else—tells us that natural causes do not give rise to complex structures such as a linguistic message.

When we find "John loves Mary" written in the sand, we assume it resulted from an intelligent cause. Experience is the basis for science as well. When we find a complex message coded into the nucleus of a cell, it is reasonable to draw the same conclusion. Science uses controlled experiments to determine what sort of results occur under given conditions. The results we observe to occur consistently and regularly are the basis of the laws we formulate.

In other words, when scientists probed the nucleus of the cell, they eventually stumbled upon a phenomenon akin to finding "John loves Mary" written in the sand, or "Vote for Smedley" written in the sky. The greatest difference is that the DNA text is much more complex. If the amount of information contained in one cell of your body were written out on a typewriter, it would fill as many books as are contained in a large library.

Are natural causes capable of producing these kinds of patterns? To say that DNA and protein arose by natural causes, as chemical evolution does, is to say complex, coded messages arose by natural causes. It is akin to saying "John loves Mary" arose from the action of the waves, or from the interaction of the grains of sand. It is like saying the painting of a sunset arose spontaneously from the atoms in the paint and canvas. When in our experience have we ever witnessed such an event? Whenever we recognize a sequence as meaningful symbols we assume it is the handiwork of some intelligent cause. We make that assumption even if we cannot decipher the symbols, as when an archaeologist discovers some ancient inscription on stone. If science is based upon experience, then science tells us the message encoded in DNA must have originated from an intelligent cause.

Even the briefest message, if meaningful, speaks unmistakably of an intelligent source.

What kind of intelligent agent was it? On its own, science cannot answer this question; it must leave it to religion and philosophy. But that should not prevent science from acknowledging evidences for an intelligent cause origin wherever they may exist. This is no different, really, than if we discovered life did result from natural causes. We still would not know, from science, if the natural cause was all that was involved, or if the ultimate explanation was beyond nature, and using the natural cause.

Genetics and Macroevolution

OVERVIEW SECTION 2

Behold the giraffe: oversized limbs, stretched-out neck, ungainly posture—everything apparently precariously out of proportion. And yet its parts are marvelously coordinated with each other; it moves with graceful ease and delivers such a powerful kick that it has few natural enemies.

The outlandish body shape of the giraffe has been a puzzle to evolutionists since before the time of Darwin. Jean Baptiste de Lamarck, one of Darwin's predecessors, suggested that the giraffe's long neck resulted from its constant stretching upward to reach leaves to eat. Bone structure changed in response to the animal's need to reach ever higher. But scientists now know that body structure does not respond to an organism's needs or habits. If it did, Olympic racers should give birth to yet faster racers, and the children of intellectuals should be even smarter than their parents.

Darwin's theory of **natural selection** turned Lamarck's explanation around: instead of the environment giving rise to habits, which in turn produce new traits, Darwin maintained that something within the organism itself gives rise to new traits, which are then either preserved or weeded out by the environment. In place of an organism needing a longer neck to survive, Darwin put an environment favoring organisms with longer necks, and then preserving any that happened along—the giraffe, as it turned out.

Clearly, it was important to Darwin's theory to locate the "something" within organisms that gives rise to new traits. To maintain that organisms either survive or perish depending on whether they are well suited to the environment was no new insight. What was unique about Darwin's theory was his idea that some force within organisms could produce new traits that over time would accumulate to produce an entirely novel sort of organism.

What force could this be? Darwin himself did not know. But ironically, at the same time Darwin was constructing his theory, an Austrian monk named Gregor Mendel was conducting experiments to answer just that question. Mendel discovered that traits could be lost in one generation only to reappear in a later generation. For example, when he crossed a pea plant bearing wrinkled seeds with one bearing round seeds, all the offspring in the first generation had round seeds. Was the wrinkled trait lost? Not at all; it reappeared in the next generation of pea plants.

Mendel concluded that heredity is governed by particles (later called genes) passed from parent to offspring. A trait might disappear temporarily, but the gene that codes for the trait remains present within the organism and may be passed on to its offspring. When a breeder, for example, causes some characteristic to appear or disappear, this represents neither a true gain nor a true loss. It represents merely the interplay of **dominant** and **recessive** genes. A "lost" trait may still be present and may reappear. A "new" trait that seems to appear out of nowhere may not be new at all but simply the expression of a recessive gene that existed all along. When breeders produce new show dogs or meatier cattle, they have actually shuffled genes around to bring some of these recessive genes to expression.

The irony is that Darwin was developing a theory of constant change at the same time Mendel was demonstrating that living things are remarkably stable. Perhaps partly because of Darwin's success in focusing attention on change, Mendel's theory was not taken seriously until the first decade of the twentieth century. Up to this point, the most commonly held concept of inheritance was a "paint-pot" model—the father's contribution blends with the mother's in the same way blue and red paint blend to form purple. Critics of Darwin pointed out that if this model were correct, any new trait that might evolve would be lost through subsequent blending, just as the original blue and red colors are lost when they form purple.

By contrast, in Mendel's model genes behave more like separate particles, being inherited essentially unchanged. When Mendel's work was re-discovered, it was welcomed enthusiastically by Darwinists and integrated into the theory of natural selection. (This modification of Darwinism is referred to as neo-Darwinism). Mendelian genetics held promise for Darwinism. For instance, it explained why a single new advantageous trait could survive and eventually become dominant in a population.

Yet Mendelian genetics has proved to be a mixed blessing for Darwinian theory. On the one hand, it provides the stability necessary for a trait to become established in a population. On the other hand, stability is just what Darwinism doesn't need if change is to be so far-ranging as to produce the whole complex web of life from a single-celled organism. How does change occur within the framework of Mendelian genetics? This is the question we will answer in the next chapter.

Does Nature Select?

Darwin bred animals and was impressed by what could be accomplished through breeding. And with good reason; by selecting animals with particular traits and allowing them to reproduce, breeders are often able to create greater differences within a single species than exist between species in the wild. Could not nature do the same and much more, Darwin asked, given enough time?

Darwin's theory begins with the observation that, in nature, many more offspring are born than survive. Since there is variation between individuals, it may happen that some offspring acquire a trait not present in their peers—say, longer legs. If this trait renders the youngsters better suited than their peers to the ecological niche they inhabit, they have a better chance of surviving, and thereby passing the trait on to their offspring. If the trend continues over several generations, eventually animals that possess longer legs will come to outnumber those who do not. The trait has become established.

Darwin dubbed this process natural selection, to emphasize its parallels to what breeders do when they select for given traits. Unfortunately, the term implied that nature was capable of actually "selecting"—of foreseeing what is needed, of choosing appropriate traits, of guiding and directing the process of evolution. Of course, nature can do none of these things. Natural selection is merely the interaction between organism and environment that weeds out harmful traits and allows helpful traits to become established.

Darwin did not originate the idea of natural selection. Several naturalists—notably Edward Blyth—had observed that natural selection takes place in nature. However, they had described it as a purely conserving force, a mechanism for weeding out unfit individuals and thereby aiding the survival of existing species. Darwin's innovation was to declare natural selection a force to produce new species.

Yet natural selection does not produce new characteristics. It only acts upon traits that already exist. The real source of evolutionary novelty—of new traits and structures—must be located in the genetic material.

How Living Things Change

People sometimes give the impression that any change is evidence for Darwinism. But Darwinism is not just any change. It is a very special kind—the transformation of one type of organism into another. Picture in your mind an evolutionary tree. The change produced by breeders is horizontal change, the flowering and elaboration of a single branch on that tree. What is needed,

Dogs are bred to develop widely differing offspring, not by adding genetic material to the gene pool, but by selecting smaller sets of genes from the larger and richer store of genetic material.

however, is vertical change leading up the evolutionary tree and creating a new branch.

To put it another way, breeders can produce sweeter corn or fatter cattle, but they have not turned corn into another kind of plant or cattle into another kind of animal. What breeders accomplish is diversification within a given type, which occurs in **microevolution**. What is needed is the origin of new types, or **macroevolution**.

Neo-Darwinism assumes that microevolution leads to macroevolution. To put that into English, it assumes that small-scale changes will gradually accumulate and produce large-scale changes. The genetic sources of change in living things are mutation and recombination.

New Patterns in Old Genes

Most variations are produced by recombination of existing genes. The tremendous differences that divide a Pekingese, a Poodle, and a Greyhound illustrate the range of variation that may exist within the **gene pool** of a single interbreeding population. These variations are produced when dog breeders isolate particular genes governing size, curly hair, or speed within a single breed. The genes can be combined and recombined in a vast number of different ways. Most changes in the living world are produced in this way—not by the introduction of anything new into the gene pool, but by simple recombination of existing genes.

Intensive breeding may produce interesting and useful varieties, but it tends to deplete the adaptive gene pool of the lineage, leading to increased susceptibility to disease or environmental change. It also tends to concentrate defective traits through inbreeding, and the farther the morphology is shifted from species norm (average), the more it produces developmental discordance, stress, and decreased fertility. Such stressed populations may often show a tendency to rebound toward the species' average morphology.

Populations seem to retain an average morphology, and the same level of variability. Chromosomal recombination may exchange parts of gene sequences. But there is no evidence that such "new" genes can provide the novel traits which natural selection needs if they are to accumulate for the endless vertical change necessary for Darwinian evolution.

Mutations: Interference in the Genetic Message

Recombination is merely reshuffling of existing genes. The only known means of introducing genuinely new genetic material into the gene pool is by mutation, a change in the DNA structure. Gene mutations occur when individual genes are altered from exposure to heat, chemicals, or radiation. Chromosome aberrations occur when sections of the DNA are duplicated, inverted, lost, or moved to another place in the DNA molecule.

As the central mechanism of evolution, mutations have been studied intensively for the past half century. The fruit fly (see Figure 3) has been the subject of many experiments because its short lifespan allows scientists to observe many generations. In addition, the flies have been bombarded with radiation to increase the rate of mutations. Scientists now have a pretty clear idea what kind of mutations can occur.

There is no evidence mutations create new structures. They merely alter existing ones. Mutations have produced, for example, crumpled, oversized, and undersized wings. They have produced double sets of wings. But they have not created a new kind of wing. Nor have they transformed the fruit fly into a new

Figure 3. *Rapid reproduction and abundant supply make fruit flies, especially Drosophila, excellent subjects for experiments designed to investigate mutations.*

kind of insect. Experiments have simply produced variations of fruit flies.

Mutations are quite rare. This is fortunate, for the vast majority are harmful, although some may be neutral. Recall that the DNA is a molecular message. A mutation is a random change in the message, akin to a typing error. Typing errors rarely improve the quality of a written message; if too many occur, they may even destroy the information contained in it. Likewise, mutations rarely improve the quality of the DNA message, and too many may even be lethal to the organism.

Macroevolution

All changes observed in the laboratory and the breeding pen are limited. They represent microevolution, not macroevolution. These limited changes do not accumulate the way Darwinian evolutionary theory requires in order to produce macro changes. The process that produces macroevolutionary changes must be different from any that geneticists have studied so far.

What is that process? Although research is underway to test various suggestions, at present there is no accepted genetic theory to replace neo-Darwinism. Evolutionists continue to be committed to the belief that macroevolution occurs but are uncertain how it occurs.

Intelligent Design: Package Deal

Let us return to the giraffe. The giraffe's long neck may appear awkward, but it is actually an integral part of the animal's overall structure. The standard explanation for the giraffe's long neck is the advantage it gives the animal when competing for food with shorter-necked varieties. This advantage would have promoted the long-necked variety's survival in greater numbers. That may be true, but the fact is that the giraffe also bends its head down to the ground to eat grass and drink water. Given the giraffe's long legs, its neck may just as well be required to reach the ground as the trees. And both the long neck and long legs facilitate feeding in tree tops. The giraffe is an

A drinking giraffe. When a giraffe bends its head to the ground to graze or drink, only an adaptational package of sophisticated blood pressure controls keeps the blood vessels in the giraffe's brain from bursting.

adaptational package in which each part is suited to the others. Trying to explain which one came first is like trying to decide which came first, the chicken or the egg.

The story doesn't end here. The giraffe requires a very special circulatory system. When standing upright, its blood pressure must be extremely high to force blood up its long neck; this, in turn, requires a very strong heart. But when the giraffe lowers its head to eat or drink, the blood rushes down and could produce such high pressure in the head that the blood vessels would burst. To counter this effect, the giraffe is equipped with a coordinated system of blood pressure controls. Pressure sensors along the neck's arteries monitor the blood pressure and activate contraction of the artery walls (along with other mechanisms) to counter the increase in pressure.

In short, the giraffe represents not a mere collection of individual traits but a package of interrelated adaptations. It is put together according to an overall design that integrates all parts into a single pattern. Where did such an adaptational package come from?

According to Darwinian theory, the giraffe evolved to its present form by the accumulation of individual, random changes preserved by natural selection. But it is difficult to explain how a random process could offer to natural selection an integrated package of adaptations, even over time. Random mutations might adequately explain change in a relatively isolated trait, such as color. But major changes, like the macroevolution of the giraffe from some other animal, would require an extensive suite of coordinated adaptations. The complex circulatory system of the giraffe must appear at the same time as its long neck or the animal will not survive. If the various elements of the circulatory system appear before the long neck, they are meaningless. If the long

Giraffe evolution

As viewed from the Far Side. Notice that in Far Side's giraffe some parts of the adaptational package fail to arrive on schedule!

legs are preceded by the long neck, the weight and accessibility of the neck without the powerful kick makes the giraffe easy prey to natural enemies. The interdependence of the structures, therefore, strongly suggests that the overall, integrated package was present from the beginning. Scientific literature often reports such interdependence of structures. In some cases, as with certain brachiopods, this interdependence traces right back to their abrupt appearance in the fossil record with the first evidences of diverse animal body plans on earth.

How likely is it that random mutations will come together and direct the formation of just one new structure? Let's say the formation of an insect wing requires only five genes (an extremely low estimate). Although it is almost certainly insufficient, let's also suppose that the new wing information required could arise from a single mutation per gene. A reasonable estimate is that, at most, a single new mutation will occur in only one individual, out of a population of 1,000. At that rate, the probability of two mutations

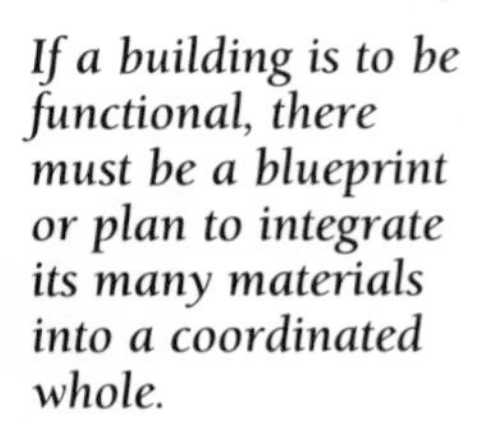

If a building is to be functional, there must be a blueprint or plan to integrate its many materials into a coordinated whole.

appearing in the same individual is one in 1,000,000. The odds of five mutations occurring are one in one thousand million million. Thus all biologists recognize there is no realistic chance that all five mutations will occur within the life cycle of a single organism. Taking large populations into account reduces the odds against these mutations, but it also increases the odds against their converging in a single mating pair. Yet, an organism is made of many structures that must appear at the same time and work together in an integrated whole, if they are not to work to its disadvantage.

To explain the appearance of new body types, a mechanism capable of generating not piecemeal change, but integrated, systemic change (affecting an organism's overall physical structure) seems to be needed. Minor changes caused by recombination of genes and by mutation may be acted upon by natural selection to fine-tune an organism, enabling it to fit better within its ecological niche. But no amount of fine-tuning of its current body plan will produce a new body plan.

In creating a new organism, as in building a new house, the blueprint comes first. We cannot build a palace by tinkering with a tool shed and adding bits of marble piecemeal here and there. We have to begin by devising a plan for the palace that coordinates all the parts into an integrated whole.

Darwinian evolution locates the origin of new organisms in material causes, the accumulation of individual traits. That is akin to saying the origin of a palace is in the bits of marble added to the tool shed. Intelligent design, by contrast, locates the origin of new organisms in an immaterial cause: in a blueprint, a plan, a pattern, devised by an intelligent agent.

The Origin of Species

OVERVIEW SECTION 3

Hawaiian Honeycreepers are among the most unusual and colorful birds found anywhere in the world. Their plumage reflects a beautiful range of colors, and the shapes of their bills vary widely too. One variety (*Hemignathus munroi*) possesses a unique adaptation: the lower bill is straight and heavy and is used like a chisel, woodpecker-style, to bore into the wood to find insects, while the upper bill is long and curved and is used as a probe to pry out insects. By contrast, Honeycreepers on the North American mainland are relatively drab birds. Why are the Hawaiian Honeycreepers so different from their counterparts on the mainland?

Some of the world's most colorful species inhabit islands. Cut off from the mainland, and exposed to new habitats, a species develops in new and unusual directions. We noted in the previous section that change is limited to variation

Hawaiian Honeycreepers. There are distinct differences between the various Hawaiian Honeycreepers, yet they probably arrived in Hawaii originally in a single variety. A smaller gene pool may well be the secret of their beauty which excels their mainland counterparts. Hemignathus munroi shown at lower right.

within existing groups of plants and animals. Yet within those boundaries, there can be rich diversity. In this chapter, we will see how such diversity may arise.

As any dog owner knows, breeds retain their distinctive characteristics only when they are prevented from interbreeding. The proud owner of a female Rhodesian Ridgeback, or Doberman Pinscher, or St. Bernard, does not let his dog breed with any mongrel in the neighborhood. For if he does, he, too, will own mongrels; in fact, a litterful of them.

In nature as well, distinctions are maintained by mechanisms that prevent interbreeding. A group may split off from its parent population and become isolated. Eventually, it may no longer interbreed with members of the parent population. This is termed *reproductive isolation* and may take place in one of the following ways.

Physical Barrier

Reproductive isolation may occur when some physical barrier cuts a group of organisms off from its parent group. Birds may be blown by storms onto an island where great distance prevents return to the mainland (which is apparently how finches reached the Galapagos Islands). A raft of vegetation may float across the ocean and deposit seeds on a distant shore ready to germinate. Animals may cross to a new region by a land bridge that later erodes away, preventing their return. Insects may somehow cross a high mountain range that keeps them from flying back freely. The redirection of a river may cut through a population and split it into two parts.

In such cases, the new group is not likely to have a full complement of genes from its parent group. Once physically isolated, the two groups may change in different directions. Eventually they may not be able to interbreed any more, even if they are brought back into contact with each other (illustrated on opposite page).

For example, there are two forms of stickleback fish in Belgium. One of them lives the year round in fresh waters, mainly in small creeks. The other lives in the sea in the winter, but migrates to river estuaries in spring and in summer, where it breeds. The two populations differ to some extent in body structure and traits, but the variations overlap broadly. Hybrids between the two forms can be obtained by artificial insemination of the eggs, but appear rarely in natural habitats. The two fish apparently derived from a single species, yet one is capable of living in salt water, the other in fresh water.

The Break-Up Of a Breeding Chain

Reproductive isolation may occur when a breeding chain is broken. A breeding chain is not unlike what we see in dogs. A Chihuahua may not breed with a Great Dane because of sheer size, yet it will breed with other dogs closer to its size. These dogs in turn will breed with other dogs slightly larger in size, until finally we reach the Great Dane. In other words, though the two extremes cannot interbreed, there are intermediate breeds connecting them. Therefore all dogs have long been considered a single species.

Such a breeding chain may exist in nature as well. A classic example is a species of butterfly living in the Central and South American rain forests. These forests originally covered an immense area and presumably were inhabited by a single species. Observations have indicated, however, that the original species has broken into various races, or subpopulations. Adjacent races can interbreed with each other but those at

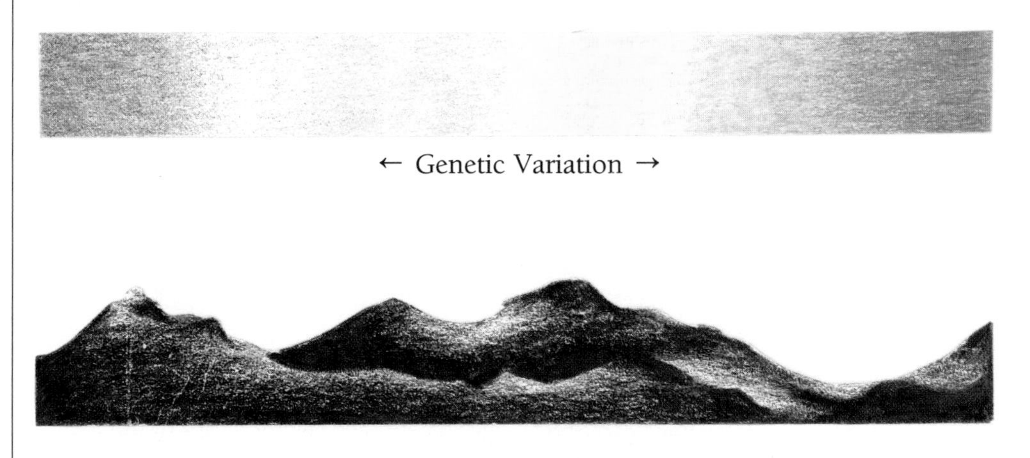

A breeding chain. A species inhabiting an extended area may develop distinct subpopulations, each interbreeding among themselves or with adjacent subpopulations.

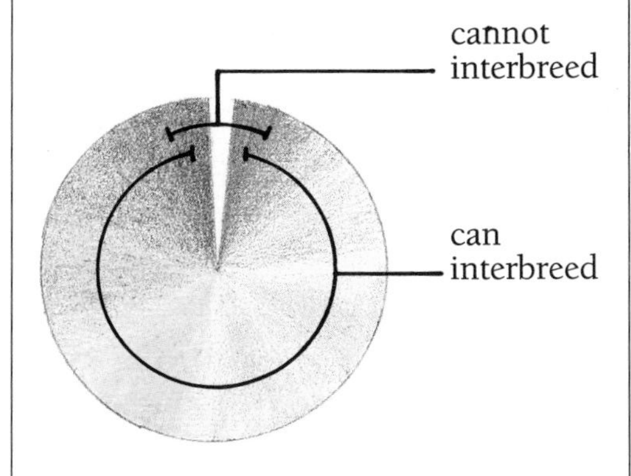

When the development of subpopulations is advanced, subpopulations on opposite ends of the breeding chain can lose their ability to interbreed.

the opposite edges of the rain forest show less interfertility.

What would happen if the intermediate links disappeared? If some environmental event, say the development of a great plain, cut the rain forest in two, the intermediate links would die out. Since the extreme subpopulations do not interbreed, they would have attained reproductive isolation.

Ecological Isolation

Organisms that remain within the same physical area may nevertheless become isolated through adaptations to fit slightly different ecological niches. This may be seen particularly in insects. The life-cycle of an insect is generally short and several generations can succeed one another within a single year. A change in the environment can thus produce rapid consequences.

A North American fruit fly (*Rhagoletis pomonella*) has divided into two populations that depend on different trees in the same area. One lives on hawthorns and the other on apple trees, and the two populations do not cross-breed. How did this happen?

Two centuries ago, there were only hawthorn flies. The females laid their eggs in August on the hawthorns and at the end of September the larvae fed on the red fruits. Later, apple trees were introduced into the area. There is enough genetic variability among fruit flies that the population gave rise to some individuals that reproduced a month early. These flies now found apple trees on which to lay their eggs, and the resulting larvae fed on apples, which ripen earlier than hawthorn fruit. A new population, dependent on the apple tree, thus became established. Because of the one-month difference in their mating schedules, the two flies do not mate in the wild. Since gene exchange no longer occurs, the two populations may eventually develop into separate species. Yet they look exactly alike and are still capable of mating in the laboratory.

Genetic Drift

Changes in the genetic structure due to reproductive isolation are more likely to occur in small populations because they are more susceptible to **genetic drift.** Small populations usually have a different ratio of genes than larger central populations. This occurs, because the laws of genetics are statistical. When a black guinea pig mates with a white one, the laws of genetics may tell us to expect a certain percentage of white offspring, and a certain percentage of black ones. But in particular matings, the offspring may be all white or all black. Why? Because, according to the laws of probability, the smaller the number of cases, the larger the chance of a deviation from the expected results.

Let's illustrate that principle. When you toss a coin, the laws of probability tell you that the chance of getting heads or tails is even. If you toss it only six times and tally the results, you may not in fact get an even ratio. A small sampling is not likely to yield the predicted outcome. Yet if you take a large sampling by tossing the coin ten thousand times, the results will come very close to an even 50:50 ratio.

In a small population, each generation is a small sampling of the genes of the previous generation. Thus, common genes may become uncommon, and the uncommon, common. Gene ratios in a small, isolated part of a population may be very different from the ratios that exist over the entire population.

Consider the following over-simplified example. Suppose that in some bird species, the **allele** "A" imparts the ability to fly, whereas the recessive allele "a" produces a grounded bird ("aa"). If birds with AA and Aa combinations arrive and breed on a small island, some flightless individuals will inevitably be hatched. If the birds have no natural enemies, the flightless variant may survive. If the population is small, and genetic drift levels are high, some generations later, the entire population could be flightless.

There is a recorded example of the potential strong effects of selection in small populations where genetic drift probably led to flightlessness. Ornithologist W. Oliver tells the story of the flightless wrens of Stephen's Island near New Zealand. In 1894 a lighthouse was placed on the island. That year the lighthouse-keeper's cat brought in several specimens of flightless wrens. No more specimens have ever been found, so apparently the cat that discovered the species also exterminated it. As biologist Tom Emmel pointed out, its limited variability "did not allow this flightless wren to cope with a change in environment—the introduction of one domestic cat."

In some cases, the new varieties that appear may no longer be able to breed with the parent population. For example, the Madeira rabbits are the descendants of ordinary domestic European hares, brought to the Madeira Islands by colonists in the late Middle Ages. Yet today, the Madeira rabbits are quite different from other European hares in both appearance and behavior, and can no longer interbreed with them.

Founder Effect and Bottlenecks

Genetic drift is particularly effective in specific cases developing out of the **founder effect**, and from population bottlenecks. A pair or handful of some species may become isolated from its parent population and establish an entire new population. They move into a new habitat or ecological niche and colonize it. Not only does the daughter population begin with a different gene ratio from the parent population, but also once it is isolated, change can occur quickly and dramatically. Biologists call this establishment by a small population the founder effect. Because members of a small group are forced to interbreed closely, a mutation can become widespread very quickly.

Take, for instance, the Amish people of Pennsylvania. Because they are descendants of only about 200 settlers, who tended to marry among themselves, the Pennsylvania Amish have a greater percentage than the American average of genes for short fingers, short stature, a sixth finger, and a certain blood disease.

The examples of the Stephen's Island wren and the Madeira rabbits may also be examples of the founder effect, where a small group—possibly even two individuals—colonized a new island habitat. Since two individuals represent a very small population, it is highly likely that the gene frequencies represented will show some deviation from those in the larger parent population.

The **bottleneck effect** is much the same sort of phenomenon. If some environmental event such as drought greatly reduces the size of a population, genetic drift will move gene frequencies quickly. Further, the gene frequencies of the surviving population may vary greatly from those in the original population as a result. As the survivors increase and establish themselves, they may be quite different from the population that preceded them, but more like each other.

Is Speciation Macroevolution?

The process we have been describing is referred to by biologists as **speciation** (the development of new species). A commonly accepted definition of species is an interbreeding population with fertile offspring. This definition is arbitrary, and taxonomists find that it is sometimes operationally inadequate for classifying organisms. Based on this definition, however, a new species is held to be a race or variety that has become reproductively isolated from its parent population. Is this macroevolution?

Many believe it is. Once reproductive isolation occurs, the road appears to lie open to large evolutionary change. Each of the two separated populations may now continue to evolve further independently.

Yet even speciation represents only limited change. Stickleback fish may diversify into fresh-water dwellers and salt-water dwellers, but both remain sticklebacks. One fruit fly may breed on apple trees and another on hawthorn trees, but both remain fruit flies. Speciation is a means of creating diversity within types of living things, but macroevolution is much more than diversity.

Macroevolution requires an increase of the gene pool, the addition of new genetic information, whereas the means to speciation discussed above represent the loss of genetic information. Both physical and ecological isolation produce varieties by cutting a small population off from its parent population and building a new group from the more limited genetic information contained in the small population. A large population carries a genetic reserve, a wealth of concealed recessive genes. In a small group cut off from the parent population, some of these recessive traits may be expressed more often. This makes for interesting diversity, but it should not blind us to the fact that the total genetic variability in the small group is reduced.

The appearance of reproductively isolated populations represents microevolution, not macroevolution. It is one of the ways in which horizontal diversification can occur. To use our earlier illustration, it is a mechanism for the flowering of a branch in the evolutionary tree, not for establishing new branches. Vertical

Pigeons bred selectively for long distance racing.

change—to a new level of complexity—requires the input of additional genetic information. Can that information—the ensembles of new genes to make wrens, rabbits, and hawthorne trees be gleaned from random mutations? Thus far, there appears to be good evidence that the roles mutations are able to play are severely restricted by and within the existing higher level blueprint of the organism's whole **genome**. To go from a one-celled organism to a human being means that information must be added to the genetic messages at each step of the way. Mechanisms for the loss of genetic information cannot be used as support for a theory requiring vast increases of genetic information.

Speciation is actually akin to what breeders do. They isolate a small group of plants or animals and force them to interbreed, cutting them off from the larger gene pool to which they belong. Centuries of breeding testifies to the fact that this produces limited change only. It does not produce the open-ended change required by Darwinian evolution.

The Fossil Record

OVERVIEW SECTION 4

It is difficult for us to imagine the surprise people felt when they first discovered fossils. To find the shape of an animal entombed in the earth, and made of stone—what mystery was this? In superstitious times, it was easy to credit fossils to some mysterious force within the earth itself. Only over time did people become convinced that once-living plants and animals could actually turn to stone if buried under the right conditions. If buried quickly enough (before being eaten away by decay or scavengers), and if buried where the groundwater is rich in the right minerals, any plant or animal can become a fossil.

Scientists now read the fossil record as a chronicle of life in former ages. Skele-

tons, footprints, leaves, spores, animal tracks, feathers, worm burrows, and even bits of hide can all be found as fossils. By interpreting these clues, scientists seek to reconstruct what living things were like in the past.

What story do the rocks tell? Like many things in science, the answer to that question depends upon one's interpretation of the facts. There are several basic features of the fossil record that must be accounted for by any interpretation.

The Cambrian Explosion

Taxonomy is the science of biological classification. Living things are classified as members of a species, then of a genus, family, order, class, and phylum. Each of these categories is referred to as a **taxon** (plural, **taxa**).

The vast majority of the known animal phyla appear as fossils, or are thought by many scientists to have arisen, within a very short period, geologically speaking. There is a virtual "explosion" of life forms recorded in the rocks at the beginning of the Cambrian. Any theory of the origin and development of life must explain how such a dramatic range of body plans made the early, abrupt appearance they did.

Gaps in the Fossil Record

Although the fossils appear to form a rough sequence, the various taxa are not connected to one another. There is no gradual series of fossils leading from fish to amphibians, or from reptiles to birds. Instead, fossil types are fully formed and functional when they first appear in the fossil record. For example, we don't find creatures that are partly fish and partly something else, leading gradually, in the dozens of characteristics which they exhibit, to today's fish. Instead, fish have all the characteristics of today's fish from the earliest known fish fossils, reptiles in the record have all the characteristics of present-day reptiles, and so on.

Stasis

Once a taxon makes its appearance in the fossil record, it remains substantially unchanged. Instead of gradually transforming into another taxon, the only change it exhibits is variation and diversification within the bounds of the original taxon. This characteristic of many species in the fossil record is referred to as **stasis**. In other words, the fossils agree with what breeders have discovered: we may produce all sorts of interesting and unusual varieties of roses or dogs, but each retains the diagnostic features that make it a rose or a dog.

Darwin: The "Gravest Objection Against My Theory"

Does Darwin's theory match the story told by the fossils? To find out, we must first ask, What kind of story would it match? His theory proposed that living things formed a continuous chain back to one or a few original forms. If the theory is true, the fossils should show a continuous chain of creatures, each taxon leading smoothly to the next. In other words, there should be a vast number of **transitional** forms connecting each taxon with the one that follows. The differences separating major groups in taxonomy (such as invertebrates and the first fish) are so great that they must have been bridged by a huge number of transitional forms. As Darwin himself noted in *The Origin of Species*, "The number of **intermediate** varieties, which have formerly existed on earth, [must] be truly enormous."

Yet this immense number of intermediates simply does not exist in the fossil record. The fossils do not reveal a string of creatures leading up to fish, or to reptiles, or to birds. Darwin conceded this fact:

> "Why then is not every geological formation and every stratum full of such intermediate links? Geology assuredly does not reveal any such finely graduated organic chain."

Indeed, this is, in Darwin's own words, "the most obvious and gravest objection which can be urged against my theory."

Yet, fossil findings were still quite patchy in Darwin's day, scattered here and there across the "map" of fossil types. Darwin predicted that intermediates connecting these individual fossils would still be found.

Thus began the search for "missing links." Convinced by Darwin's theory that fossil taxa must be linked by a graded series of intermediates, scientists began an intensive search. A few odd-ball types did show up that failed to fit neatly within any existing taxa, like Archaeopteryx (ar-kee-OP-tuh-riks, a proposed transition between reptiles and birds), which were initially hailed as transitional. The puzzle raised by Archaeopteryx has to do with the "avian complex" or adaptational package of characteristics making flight possible in birds. The feathers in Archaeopteryx are identical to those in modern birds, having the structure of a genuine airfoil. Yet in place of the "avian complex," Archaeopteryx has eight reptilian features. No process capable of sculpting its feathers while leaving its other reptilian features untouched is known to current Darwinian theory. In fact, Archaeopteryx has only one bird-like feature, much like the duck-billed platypus (see Figure 4) living in Australia today. The platypus has a bill like a duck and fur like a mammal, but has never been considered transitional. Most candidates for missing-link status have fallen by the wayside.

For over a century, paleontologists (scientists who study fossils) were puzzled by this glaring lack of transitional fossils. It was one thing to hold out the hope

Figure 4. *A duck-billed platypus.*

Charles Darwin. An artist's portrait of Darwin at 40, a decade before he was to publish his book, The Origin of Species.

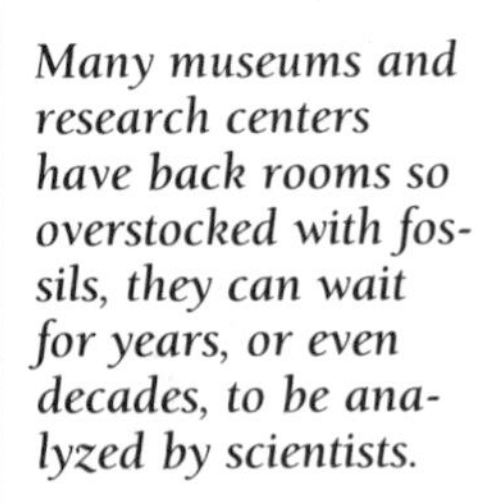

Many museums and research centers have back rooms so overstocked with fossils, they can wait for years, or even decades, to be analyzed by scientists.

of finding missing links in Darwin's day when the science of paleontology was still in its infancy. Maybe scientists had not yet searched long enough. But today, more than a century and a quarter later, few would still try to defend this explanation. Today the number of fossils that have been unearthed is staggering, and new ones are being discovered faster than they can be catalogued.

As fossil finds grew, it became apparent that the fossils were falling into a definite pattern. Instead of forming a graded series, as Darwin had expected, the fossils filled in existing taxa, leaving the gaps between them conspicuously empty. The pattern in the fossils is not a continuous chain but clusters separated by gaps. Perhaps that should not be surprising—it is, after all, the same pattern we see among living organisms today. There are many breeds of horses, but they are clearly separated from cattle; there are many varieties of corn, but no one would confuse them with wheat. Varieties cluster around a basic morphological pattern ("morphe" means form, shape, structure) rather than leading smoothly from one form to the next.

Evolution By Leaps

In recent years, many evolutionists have come to question more and more aspects of Darwin's theory. Does that mean these scientists have rejected Darwinism itself? Not at all. Most are simply seeking to formulate new ideas about how macroevolution works. Since the fossil record shows clusters separated by gaps, these evolutionary scientists are trying to conjecture what sort of process would leave such a pattern. A current view is that evolution is a step-wise process, remaining on a single level for some time (producing stasis), then jumping to the next level suddenly, thereby leaving little fossil evidence (producing gaps). By sudden is meant geologically sudden—from a few hundred to tens of thousands of years.

A major proponent of the new theory of evolution is Stephen Jay Gould, a paleontologist at Harvard University. He calls the theory **punctuated equilibrium.** "Equilibrium" refers to the fact that species remain relatively stable for long periods of time (stasis). Changes that do occur represent variation within a given taxon, not progression to a new taxon. This stability is sometimes broken—punctuated—by bursts of macroevolutionary change leading to the rise of a new taxon. Since rapid change is more likely to occur in small populations (as we saw in Section 3), the number of organisms involved in this dramatic evolutionary leap is generally thought to be few. Phrased differently, the number of transitional forms would be small. And the number that happened to fossilize would be even smaller, whence this theory's account of the gaps in the fossil record.

Punctuated equilibrium is one of several approaches currently being investigated by scientists. The attractive feature of these new approaches is that they offer an explanation for the existence of gaps in the fossil record, which is where Darwin's prediction failed. Until recently, most people viewed Darwinism as synonymous with evolution. But today scientists distinguish the inference that living things evolved from earlier forms from theories about how they evolved. Darwin's view was that macroevolution occurred in a slow and gradual manner. Gould and some other contemporary scientists believe it happened in a sudden, jerky manner.

Taking the Rocks At Face Value

The newer forms of evolutionary theory may fit the facts better, but their weakness is that they are based upon negative evidence. There is still no positive fossil evidence for evolutionary descent from one taxon to the next. Although ingenious, punctuated equilibrium advances an explanation for macroevolution's lack of evidence.

Many scientists conclude that there never was a progression from one cluster to another—that each really did originate independently. This idea accords with the theory of intelligent design. Design theories suggest that various forms of life began with their distinctive features already intact: fish with fins and scales, birds with feathers and wings, mammals with fur and mammary glands. Proponents of intelligent design face problems of their own, but feel compelled to ask, Might not gaps exist in the fossil record, not because large numbers of transitional forms mysteriously failed to fossilize, but because they never existed?

The standard Darwinian interpretation is that fossils around the world were laid down in rock strata over vast ages. Organisms that appear as fossils in lower strata lived earlier than those in higher strata. The Darwinist concludes from this that the ones in the lower strata evolved into the ones in the higher strata.

This conclusion must be drawn, however, in the absence of empirical evidence of a chain of fossils leading from lower organisms to higher ones. It is a conclusion shaped as much by philosophical commitments as evidence. If we see one organism followed by another, *and we assume that only natural causes were at work*, then we really have no choice but to conclude that the earlier organism evolved into the later one.

There is, however, another possibility science leaves open to us, one based on sound inferences from the experience of our senses. It is the possibility that an

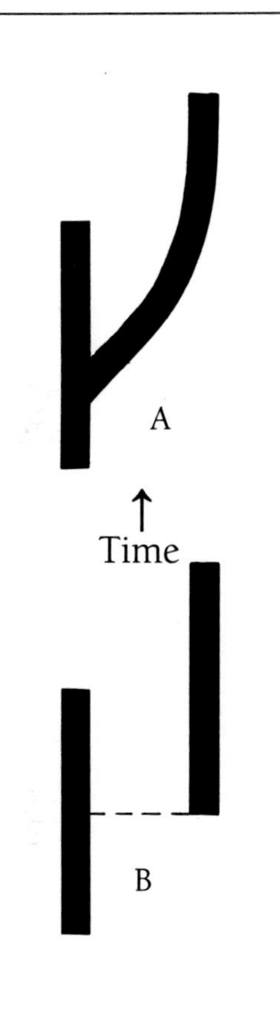

The branching process as viewed by traditional evolutionists (A) and by proponents of punctuated equilibrium (B).

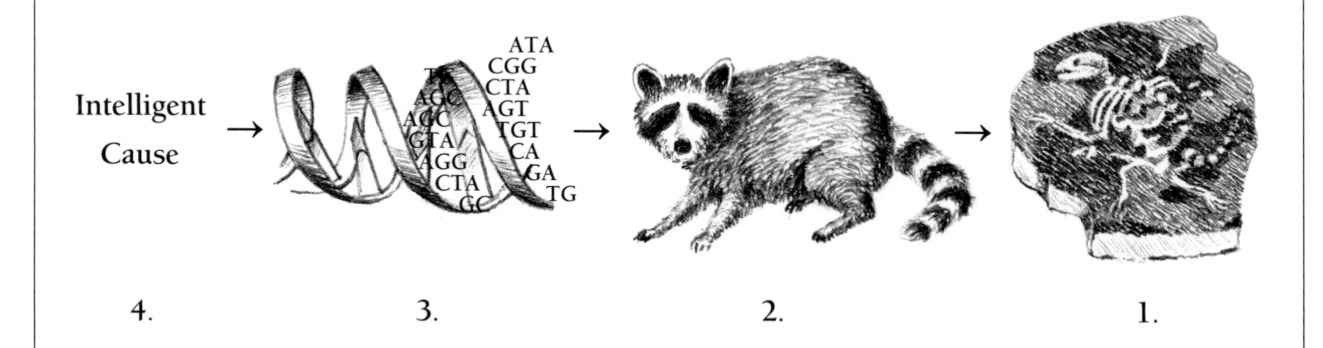

Figure 5. ***When we infer that an intelligent cause was responsible for fossils we observe, we do so through the experience of our senses.***

intelligent cause made fully-formed and functional creatures, which later left their traces in the rocks. We simply work backwards from the fossil to the creature to message text in DNA, to the intelligent cause (see Figure 5). We are free to take the evidence where it leads. If there is evidence for natural cause, then we conclude descent. If there is evidence for intelligent cause, then we conclude design. On both sides, the decision one ultimately makes regarding the fossils rests on philosophical commitments as well as on empirical data.

Fact Versus Interpretation

The fossil record is sometimes treated as the Darwinist's trump card. Despite controversies over how evolution occurred, many argue, the fossil record nonetheless establishes that it occurred. It documents that macroevolution is a fact. This argument assumes, however, that Darwinian evolution is the only reasonable interpretation of the fossil record. Yet, the intelligent design hypothesis is also reasonable. It fits well with the empirical data, with the fact that fully-formed organisms appear all at once, separated by distinct gaps. It is a mistake to claim for macroevolution the status of fact. The existence of fossils with enormous variety is a fact, and so are the changes in the distribution of those fossils over time; to read an evolutionary history of life on earth from the fossils, on the other hand, is to construct a theory. To read intelligent design from the fossils is also to construct a theory. So both Darwinism and design must take their places as theories to be considered and evaluated. That, of course, is what Darwin had in mind when he looked with confidence to a time when students "will be able to view both sides of a question with impartiality."

Homology

OVERVIEW SECTION 5

To organize living things by whether they sport fur or feather or fin—that is the job of the taxonomist. His goal is to group organisms by their similarities and to distinguish them by their differences. But taxonomy has an interpretative side as well. Once patterns of similarities have been described, the taxonomist seeks to explain what those patterns mean. Why can we classify living things in distinct categories as representatives of a species, genus, family, order, class, and phylum? Why are all vertebrates constructed on essentially the same body plan, despite the many obvious differences that separate them? Why can organisms be classified at all? Theoretically, living things could fall into a random pattern, or into shifting, overlapping groups, instead of falling into a neat group-within-group arrangement, as they do.

Mary Had a Little Lamb (*Ovis aries*)

The fact that we can classify living things at all means that we perceive degrees of similarity among them. A dog is more like a wolf than it is like a fox; as a result, the dog and the wolf are classified in the same genus (*Canis*) and the fox is classified in a different genus. Yet a dog is more like a fox than it is like a cat; so they are classified in the same family (Canidae) and the cat is classified in a different family. But a dog is more like a cat than it is like a horse; they are placed in the same order (Carnivora), and the horse is placed in a different order. Still, a dog is more like a horse than it is like a fish; therefore, they share the same class (Mammal) and the fish is in a different class. But a dog is more like a fish than it is like a worm; both dog and fish belong to a single subphylum (vertebrates) and the worm belongs to a different phylum. The dog has more in common with a worm, however, than it has with an oak tree; therefore, they are in the same kingdom (animals) and the tree is in a different kingdom (plants).

For Darwin, similarity was a major argument for evolution. He interpreted similarity as "family resemblance": two organisms are similar because they are descendants from a common ancestor. Imagine a photograph of a large extended family. The family features are obvious; brothers and sisters resemble one another most closely, cousins somewhat less, and so on. In a comparable way, say evolutionists, degrees of similarity reveal how closely organisms are related. The fact that all mammals are built on a common

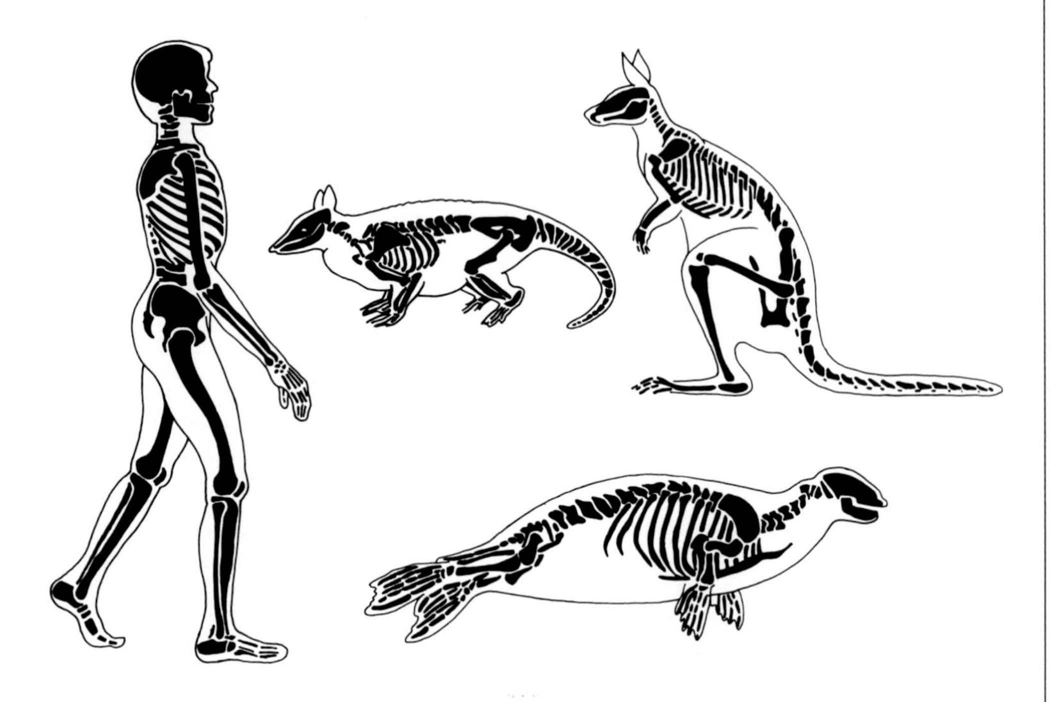

Figure 6. ***Examples of the mammalian body plan.***

body plan (see Figure 6), for instance, means they descended from a common ancestor that originally evolved that body plan. Differences among mammals reveal how the basic plan has been adapted in each species under the pressure of natural selection.

The argument from similarity has achieved even greater importance today than in Darwin's time because of the gaps in the fossil record. Because paleontologists have not been able to trace a line of descent in the fossils from one organism to another, they must rely on similarities alone to construct proposed relationships.

Patchwork Pattern

According to Darwinian theory, the larger the number of similarities between two organisms, the closer is their evolutionary relationship. Yet to discern and interpret similarities is not as simple as it may sound. Once we proceed beyond the rather obvious similarities—most birds have feathers, most fish have scales—it is not always easy to decide which organisms should be classified together. Similarities appear in a patchwork pattern that makes classification difficult.

Contradictory Similarities

Consider the marsupials—mammals that nurture their newborn in a pouch on the mother's belly (in contrast to placental mammals, such as humans). Marsupials and placental mammals are sometimes strikingly similar (see Figure 7). For instance, in skeletal structure, the North American wolf and the now-extinct Tasmanian wolf are very close—in some features, nearly indistinguishable. The

Figure 7. *Examples of the many dramatic parallels in placental (above) and marsupial (below) mammals.*

behavior and life-style of the Tasmanian wolf was likewise quite similar to that of the North American wolf. Despite these close parallels, because the two animals differ in a few features, such as their jaws, dentition, and their mode of reproduction, the standard taxonomic approach is to classify them in widely different categories—the North American wolf with the dog and the Tasmanian wolf with the kangaroo.

Besides wolves, there are also marsupial look-alikes to cats, squirrels, ground hogs, anteaters, moles, and mice. Marsupials raise an interesting question for taxonomy: If similarity is the basis for classification, what shall we do when similarities conflict? The marsupial wolf is strikingly similar to the placental wolf in most features, yet it is like the kangaroo in one significant feature. Upon which similarity do we build our classification scheme?

Function Versus Structure

Both a bird's wing and an insect's wing are used for flying. Both function in the same way: air currents pushing against the surface of the wings provide lift, and flapping the wings provides forward thrust. Yet the internal structure of a bird's wing is very different from that of an insect's wing. The bird's wing consists of flesh, supplied with food and oxygen by a network of blood vessels. Its support is on the inside, in its bones. The insect's wing, on the other hand, has no bones or blood vessels. It consists of a thin membrane stretched tightly around a network of wiry structures, similar to a kite.

Which is relevant for classification—similarity in function or similarity in structure? The first great taxonomist, Linnaeus, faced this problem and chose the latter: he classified the flying insects with other insects, because of the similarity in structure, and not with birds, which seems quite right to us as well. Similarity in structure he referred to as **homology**. Similarity in function he referred to as **analogy**.

Darwinian theorists assume that homologies are evidence of evolutionary descent. For instance, the similarity in skeletal structure among vertebrates is interpreted as evidence that all descended from a common ancestral form. Analogies, on the other hand, are held to be the result of **convergent evolution**, parallel adaptations to similar environments. The evolution of flight in insects and birds occurred by independent processes that came together (converged) in response to environmental pressures.

Yet, the distinction between homology and analogy is not always easy to draw. Structures considered homologous may be quite different in both appearance and function. Think of a bat's wing and a horse's foreleg. On the other hand, very similar-looking structures that perform similar functions may be considered merely analogous and hence irrelevant for classification. Think of the body shape of the fish and the whale. Early in his career, Linnaeus classed the Cetaceans (whales) as fish, not realizing that their fish-like shape was not a homologous, but an analogous resemblance.

Repeatedly throughout the history of taxonomy, structures of astonishing similarity at first regarded as homologous were later determined to be analogous. It would seem that many structures are mixed, being both homologous and analogous.

The Puzzling Panda

The giant panda and the lesser, or red panda present us with an eloquent illustration of the problem with homologous and analogous structures. Both pandas are

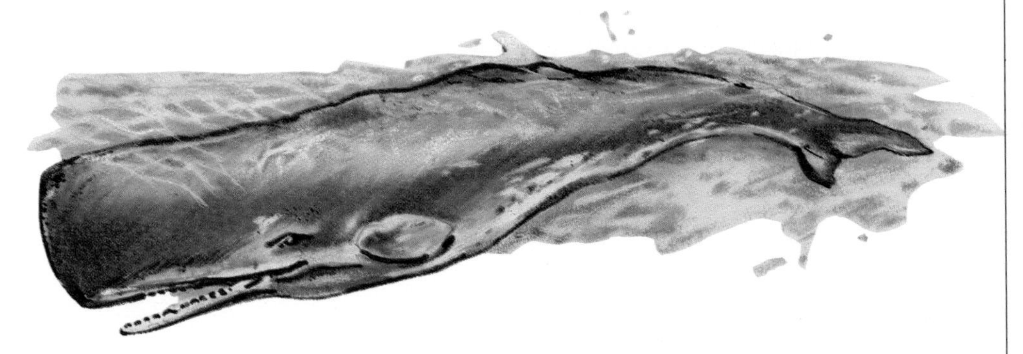

The fish-like shape of the whales is an example of similarity that does not come from close kinship. It is therefore held to be an analogous resemblance.

The lesser, or red panda. Like the giant panda, the red panda is also a bamboo-eating native of southwest China.

native to the bamboo forests of southwest China. For over a century, scientists were unable to agree whether the two pandas are members of the bear family or the raccoon family. About half the studies done on the pandas concluded that they are bears; half concluded that they are raccoons. In 1964 a study was done that is now generally accepted as the definitive interpretation. Its conclusion? That the giant panda is a bear, but the red panda is a raccoon!

Here is a classic case of scientists being unable to decide which similarities to treat as decisive. Until the 1964 study, they were unanimous on at least one point: that the two pandas are close relatives and should be classified in the same family. There are compelling similarities in skeletal structure, internal organs, behavior, and chromosome count that connect the two animals and set them off from other animals.

Perhaps the most dramatic similarity is that each panda is graced with a "thumb" (see Figure 8). It is not a true thumb but rather an enlarged bone of the wrist. Yet it operates much like a thumb and is even partly opposable. The giant panda uses it to strip bamboo, an activity in which it is engaged most of the day. The red panda's thumb is smaller than the giant panda's but is used in much the same way.

For over a century, the striking similarities between the two pandas were

Figure 8. *The panda's paw with "thumb" showing.*

considered "homologies" and the animals were classified together. Today, however, they are classified in different families and so the similarities must now be considered "analogies," which is awkward for something so peculiar as the unusual thumb.

Similarities are not always easy to discern or interpret. A given animal may resemble one group in certain features and another group in other features (as exemplified by the marsupials), requiring taxonomists to select which similarity to use in classification. On what basis does a scientist decide which set of similarities is relevant for classification (homology) and which should be ignored (analogy)?

Taxonomists have sought to devise criteria to distinguish homologies from analogies. The common view is that natural selection is more likely to produce general structures—like skeletal structure shared by the North American wolf and the Tasmanian wolf—than it is to produce highly specific structures —like marsupial pouches. Therefore, the general skeletal similarities are more reasonably explained by convergent evolution, whereas the marsupial pouches are more reasonably explained by evolutionary descent. Yet even this criterion involves subjective judgments about what evolution is most likely to have done, and taxonomists still find themselves embroiled in frequent differences of opinion over how to classify organisms. It has been said that there are as many classification schemes as there are taxonomists. While this is an exaggeration, there is some truth to the idea.

Moreover, this criterion assumes that macroevolution happened, rather than seeking to find out whether it happened or not. Unfortunately, the very term "homology" is often defined to include the concept of evolution. Most biology books today define homology as correspondence of structure derived from a common ancestor. As a result, evolutionists sometimes fall unwittingly into a circular argument: the concept of evolutionary descent is employed to explain similar structures, and then the existence of similar structures is cited as evidence that macroevolution has occurred.

The Products of Design

If the sheer fact that living things can be classified leads inevitably to a Darwinian conclusion, it is surprising that for over two millennia classification didn't have that effect. Classification went on quite successfully before the appearance of Charles Darwin in the 19th Century without employing the concept of family relationships. Instead, structures held in common by large groups of organisms were interpreted as the outworking and adaptation of an original plan.

Many things can be classified that are not derived from a common ancestor—things like cars and paintings and carpenter's tools; in short, human artifacts. What makes all Fords look similar, or all Rembrandts, or all screwdrivers, is that they are derived from a common design or pattern in the mind of the person making them. In our own experience we know that when people design things—such as car engines—they begin with one basic concept and adapt it to different ends. As much as possible, designers seek to piggyback on existing patterns and concepts instead of starting from scratch. Our experience of how human minds work provides an indication of how a primeval intellect might have worked.

A Living Mosaic

The theories of intelligent design and natural descent both have an explanation for why living things share common structures. And since both theories are able to account for similarities, the sheer existence of similarities cannot count as evidence for or against either theory. Yet there is more to consider, and that is the erratic, patchwork pattern of similarities.

Recall the puzzle of the marsupials. According to Darwinian theory, the pattern for wolves, cats, squirrels, ground hogs, anteaters, moles, and mice each evolved twice: once in placental mammals and again, totally independently, in marsupials. This amounts to the astonishing claim that a random, undirected process of mutation and natural selection somehow hit upon identical features several times in widely separated organisms.

Or take the problem of flight. The capacity for powered flight requires a tremendously complex set of adaptations, affecting virtually every organ of the body. Yet Darwinists insist that flight has evolved independently not once but four times: in birds, in insects, in mammals (bats), and in pterasaurs (extinct flying reptiles).

What such examples reveal is that similarities do not trace a simple branching pattern suggestive of evolutionary (genealogical) descent. Instead, they occur in a complex mosaic or modular pattern. Similar structures, like the hemoglobin molecule, appear here and there in the mosaic of living things, like a silver thread weaving in and out of a tapestry. Similarities may also be described as fixed patterns or discrete blocks that can be assembled in various patterns, not unlike subroutines in a computer program. Genetic programs each incorporate a different application of these subroutines, generating the diversity of biological forms we see today.

To use another analogy, similarities among living things are like preassembled units that can be plugged into a complex electronics circuit. They can be varied according to an organism's need to perform particular functions in air or water or on land. Organisms are mosaics made up from such units at each biological level. In this view, the possession of similar structures implies nothing of evolutionary ancestry.

Not only the wings, but the skeleton, the heart, the respiratory system, and many other parts of the bird are attuned to its ability to fly.

Biochemical Similarities

OVERVIEW SECTION 6

Many people think of proteins as one of the major food groups or the mainstay of a healthy diet; it would hardly occur to them that they are tiny machines. Yet, we have come to recognize proteins as just that, functioning in an astonishing variety of specific ways to build living tissue and carry out the reactions necessary for life. A typical cell can easily contain thousands of different proteins, each unique among its fellow proteins. Proteins known as repressor molecules, for example, lock up or switch off the operation of parts of the DNA, until needed. Skin is composed largely of a protein called collagen. The proteins myosin and actin in muscle cells are largely responsible for their ability to contract, and when light contacts the retina of your eye, it interacts first with a protein called rhodopsin.

These and thousands of other functions are all critically dependent upon an extraordinary integration and cooperation among the various parts of the proteins, a cooperation that far exceeds the level exhibited in man-made machines. These cooperative activities, in turn, are dependent upon the sequence of amino acids along the chain of the protein molecule.

The study of living things at the molecular level is a relatively new field. The information that scientists derive from molecular biology may be used to compare and categorize organisms, a field known as *biochemical taxonomy.* Biochemical analysis holds out the promise of making taxonomy a more precise science, because it allows differences between various organisms to be quantified and measured.

Homology Writ Small

We all intuitively regard a horse as more similar in general structure to a cow than to a bird, but there is no way of measuring the difference between them in mathematical terms. Deciding which organisms should be classed together based on comparative anatomy and homology is always plagued by an element of subjective judgment. The revolution in molecular biology changes all that. It provides a new way to compare organisms based on the structure of their proteins and DNA.

One of the important procedures of biochemical taxonomy is the determination of amino acid sequences in protein, and the sequences of triplets in DNA. Researchers employ DNA and protein sequence analyzers to determine these sequences. Many proteins are used in a variety of organisms. It has been found

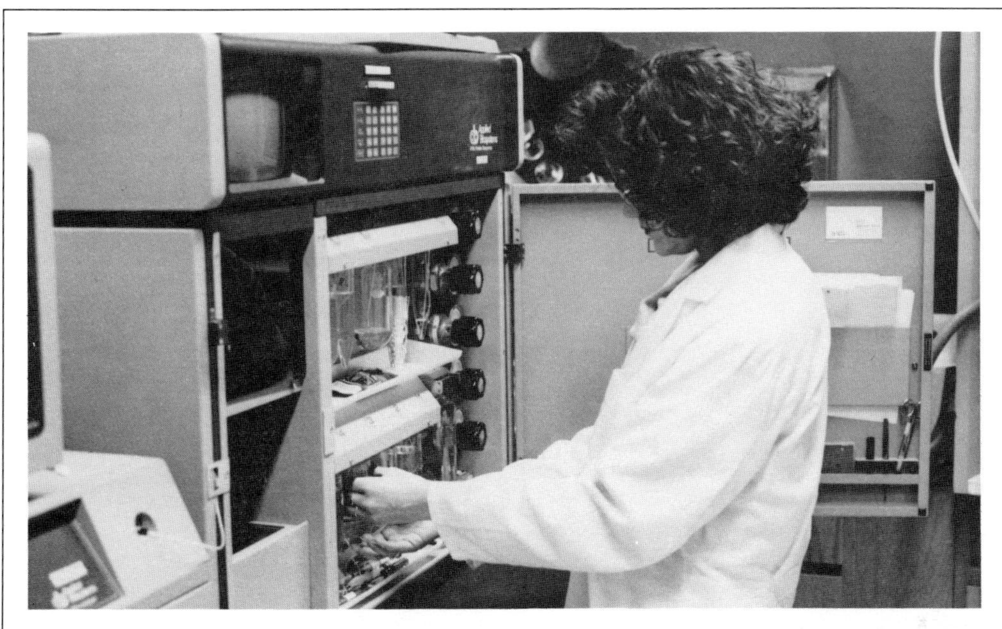

that the sequence of a given protein, say cytochrome *c*, is not fixed but varies from species to species. Usually cytochrome *c* is composed of a string of one hundred and four amino acids. Though it performs the same function and is similar enough to be recognized as the same protein, it nevertheless differs among the various taxa. The amino acid sequences from two different organisms can be compared by aligning the two sequences and counting the number of individual amino acids that differ. Similar comparisons can be made between two strands of DNA. As an illustration, consider the two series of letters below:

A	B	C	D	E	F	G	H	I	J
A	D	C	D	J	F	B	H	I	J
	1			2		3			

These two series consist of ten letters each and differ in the three positions numbered. The measure of difference is therefore 30 percent. If the series diverged in only one position, the difference would be 10 percent; if they diverged in two positions, the difference would be 20 percent, and so on.

Animals with a greater number of similarities in DNA or amino acid sequence are classified more closely taxonomically. The classification system that emerges from molecular biology to a large extent confirms classifications traditionally made by taxonomists from anatomy. That is, a horse is more like a cow than it is like a bird not only in obvious appearance but also in the sequence of amino acids of its proteins, and of triplets in its DNA.

Resemblance Revisited

By providing a new way to identify similarities, biochemists have expected to build our understanding of what those similarities mean. As we saw in Section 5, the standard Darwinian interpretation is that similarity indicates descent from a common ancestor. The greater the resemblance between two organisms, say Darwinists, the more recently they diverged from a single line of descent.

Adherents to Darwinism have hailed the findings of biochemistry as important new evidence for the theory. The fact that similarities in biochemistry parallel similarities in anatomy is held as confir-

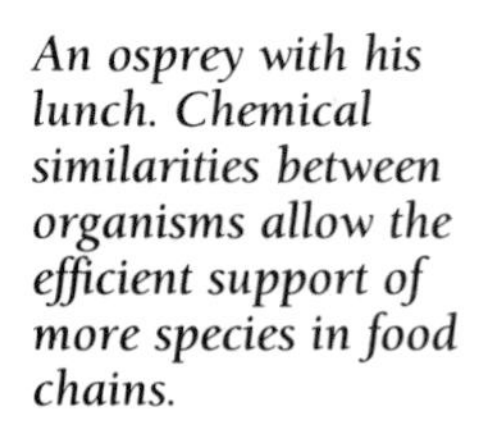

An osprey with his lunch. Chemical similarities between organisms allow the efficient support of more species in food chains.

mation of the evolutionary relationships inferred from those similarities. The proteins of humans more closely resemble those of monkeys than those of turtles, for example. This is taken as confirmation that humans share a common ancestry with monkeys.

Proponents of intelligent design read similarity in structure as a reflection of similarity in function. All living organisms must survive in the same universe and must fit into its ecological web. All must fit into a food chain. The need to function within a common universe puts common physical and chemical requirements on all organisms. It would be both logical and efficient for an intelligent agent to design living things with a common biochemical base. By the same token, it should not surprise us that organisms share similarities on both anatomical and biochemical levels. The genuinely new light biochemistry brings to the subject lies elsewhere, as we shall see.

A New Pattern

Scientists are attempting to make additional evolutionary trees, through biochemical comparisons, to check the older ones. But when measurements of the similarities between proteins are put side by side, the pattern that emerges contradicts the expectations based on Darwinism. Let's look at this pattern in detail. Table 1 shows the percent of difference in amino acid sequence in cytochrome *c* between several organisms. (Note that even when the percentages are identical for more than one organism, the actual amino acid positions where they diverge are not likely to be the same.)

	1.	2.	3.	4.	5.	6.	7.	8.	9.	10.	11.	12.	13.	14.	15.	16.	17.
1. Humans	0	1	10	12	9	10	13	13	14	17	20	17	23	19	29	38	65
2. Rhesus Monkey		0	9	11	8	11	12	12	13	16	20	17	22	19	28	38	64
3. Pig, Bovine, Sheep			0	3	4	6	9	10	9	11	16	11	15	13	25	40	64
4. Horse				0	6	7	11	12	11	13	18	13	16	15	27	41	64
5. Rabbit					0	6	8	8	9	11	16	13	16	16	24	39	64
6. Kangaroo						0	12	10	11	13	17	13	19	16	26	42	66
7. Chicken, Turkey							0	2	8	11	16	14	18	17	26	41	64
8. Penguin								0	8	12	17	14	19	18	25	41	64
9. Snapping Turtle									0	10	17	13	18	18	26	41	64
10. Bullfrog										0	14	13	19	20	27	43	65
11. Tuna											0	8	19	18	30	44	65
12. Carp												0	14	12	25	42	64
13. Dogfish													0	16	30	44	65
14. Lamprey														0	30	46	66
15. Silkworm Moth															0	40	65
16. Wheat																0	66
17. Rhodospirillum rubrum c_2																	0

Table 1. *Comparisons of the cytochrome c molecules of 17 organisms.*

Taking the line marked "humans" and tracing across, we see that the differences in sequence become greater the farther away we move on the taxonomic scale. From human to Rhesus monkey is only one percent divergence; from human to pig is ten percent; from human to a fish (carp) is 17 percent; from human to an insect (silkworm moth) is 29 percent; and so on. This finding is not surprising since it corroborates traditional taxonomic categories.

Now look at the entry for silkworm moth (No. 15 at the top of the table) and this time go down the table from vertebrate class to vertebrate class. Notice that the cytochrome *c* of this insect exhibits the same degree of difference from organisms as diverse as human, penguin, snapping turtle, tuna, and lamprey. Considering the enormous variation represented by these organisms, it is astonishing that they all differ from the silkworm moth by almost exactly the same percent.

The reason this finding is so surprising is that it contradicts the Darwinian expectation. As we move up the scale of evolution from the silkworm moth, that expectation (although it was probably never stated as a prediction) was to find progressively more divergence on the molecular level. This expectation holds true, even though we are comparing what may be described as contemporary representatives of progressively appearing classes, rather than descendants and ancestors. For example, Darwinists may consider the bullfrog, an amphibian, to be the product of several branching events that occurred after the branching of the amphibian lineage leading to reptiles. This would mean that the bullfrog, although a member of a vertebrate *order* that is ancestral to the reptiles, is not directly ancestral to them *itself*. The expectation is still that a "tree" pattern (albeit a general one) would be seen when comparing ancestral orders via just such living, non-ancestral representatives. Indeed, when comparing living organisms, Darwinism would predict a greater molecular distance from the insect to the amphibian than to the living fish, greater distance still to the reptile, and greater than that to the mammal. Yet this pattern is not found.

Consider another example. In the evolutionary scenario, fish evolved into amphibians. Thus one might expect analysis to reveal that the cytochromes in fish are most similar to the cytochromes in amphibians. But that is not the case. Figure 9 gives the percent sequence difference between cytochrome *c* in fish (carp) and several land-dwelling vertebrates.

When we compare the cytochrome *c* sequence in a variety of vertebrates, we find them to be equidistant—the same distance—from the fish; again, the expected evolutionary tree is missing. Amphibians, represented in the chart by the bullfrog, are traditionally considered closest to fish in the evolutionary scale. Yet on a molecular level they are no closer to fish than to reptiles or mammals.

To use the classic Darwinian scenario, amphibians are intermediate between fish and the other land-dwelling vertebrates. Analysis of their amino acids should place amphibians in an approximately intermediate position, but it does not. This is true for nearly every species of amphibian we choose for comparison. Based upon the evolutionary series, we would expect some amphibians to be closer to fish ("primitive" species) and others to be closer to reptiles ("advanced" species). Moreover, we might expect the horse to be the most distant. But this is not the case either. (The fact that the protein in horses, rabbits, chickens, turtles, and bullfrogs is equidistant from that in carp [see Figure 9] does not mean the former are all identical. These proteins differ from each other as well as from those of the carp. The 13 or 14 positions where their amino acids diverge from those of the carp are not the same from organism to organism.)

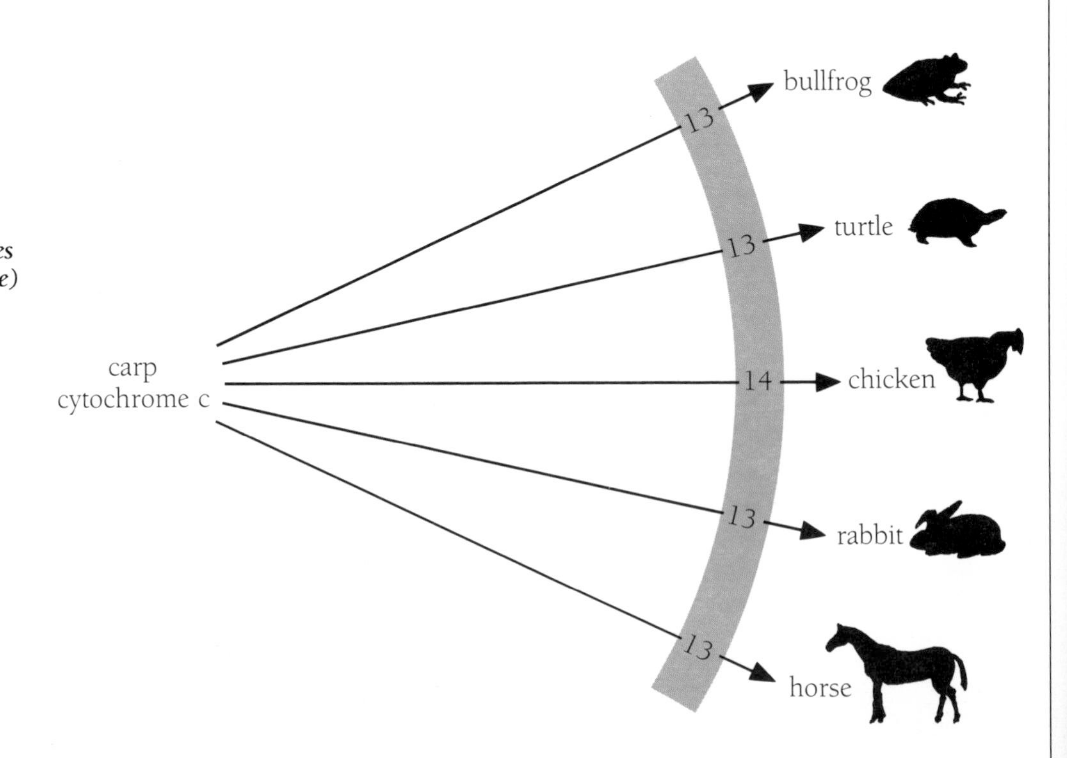

Figure 9. ***The sequence differences (percent divergence) between the cytochrome c molecules of the carp and several other vertebrates. Notice how close these percentages are to each other.***

The Molecular Clock

Some scientists have suggested that the idea of a *molecular clock* solves the mystery. The explanation they advance is that there is a uniform rate of mutation over time, so, quite naturally, species that branched off from a common ancestor at the same time in the past will now have the same degree of divergence in their molecular sequences. There are some serious shortcomings, however, with this explanation. First, mutation rates are thought to relate to generation times, with the mutation rates for various molecules being the same for each generation. The problem comes when one compares two species of the same taxon—say two mammals—with very different generation times. Mice, for instance, go through four to five reproductive cycles a year. The number of mutations, therefore, would be dramatically higher than, say, those of an elephant. Thus, they should not reflect similar percent sequence divergences for comparable proteins.

Besides that, the rates of mutation are different for different proteins even of the same species. That means that for the molecular clock idea to be correct, there must be not one molecular clock, but thousands.

The Task of Taxonomy

Taxonomy is often a tedious and thankless job. Researchers may go to great lengths to decide on a single classification. But the task of taxonomy is not only to work on the level of such minute detail; it is also to seek to paint an overall picture of the pattern of organic life.

The picture painted by Darwinists is of a continuous series of organisms. Life unfolds in a sequence of stages from simple to complex. Within each classification are some organisms that are "primitive," barely emerging from a previous, simpler stage, and others that are "advanced," on their way to a more complex stage.

The picture observed by proponents of intelligent design is groupings of related organisms within higher level categories that are independent of one another. Since proponents of intelligent design see no satisfactory evidence that all groups are derived from others, unbroken sequence is not necessary. Instead, major taxa are separate and marked, easily distinguished from others by their characteristic, integrated structures.

Which picture best fits the world revealed to us by science? Contemporary organisms can be reasonably explained on the basis of intelligent design. If we consider the family *Felidae* (cats) we see many variations, from domesticated Siamese cats to African lions and Bengal tigers. However, we readily distinguish between them and, for example, foxes, weasels, or dogs. The living world shows us a picture of clusters of organisms, each grouped around a set of defining characteristics. At the boundaries, particularly at the levels of genus and species, these distinctions are not always so clear, and may be subject to later revisions by taxonomists.

The fossil record also reveals a pattern of clustered organisms. A revolution in science has been underway over the past twenty years, with many paleontologists modifying Darwin's basic tenet of gradual change to the concept of stasis (no change) alternating with episodes of rapid change. This modification of the theory of evolution has been called punctuated equilibrium. These scientists have come to realize, after more than a century of research into the fossil world, that the pattern of fossil organisms is not, by and large, a graded series, but clusters separated by gaps.

The new data from biochemistry where organismal differences can be measured somewhat more quantitatively, generally confirm this pattern of clustering. By studying sequences of amino acids in proteins, it has been found that organisms cannot be lined up in a series, $A \rightarrow B \rightarrow C$, where A is an ancestor of B and B an ancestor of C, but are instead, approximately equidistant from most other organisms in a different taxon. This feature remains reasonably consistent over a wide range of species.

The data from a variety of fields have come together like pieces in a jigsaw puzzle. While many pieces are still missing, there is nevertheless a picture of clusters of organisms, each harboring variation in nondefining traits, and each major grouping separated from others by distinct gaps. Major advances in molecular biology have given us new, quantifiable data on the similarities and differences in living things. We must never give the impression that our present scientific knowledge has provided all the answers, but we can say that the data have not served to support a picture of the organic world consistent with Darwinian evolution.

EXCURSION CHAPTER I

The Origin of Life

Introduction

Where did life come from? That question is both ancient and up-to-date. The answer affects not only our view of the world and what we look to for meaning in our lives, but how we view ourselves as well.[1] In fact, since we all care very much about who we really are, it might well be one of the most important questions we will ever try to answer.

How life originated on earth is not only an important question to ask, it is also one of the toughest to answer. Because science had no observers stationed nearby during life's origin, it must study the question through indirect means. As a result, we can only give probable explanations. We must observe the present world and the life it contains to find clues to what happened in the past. Then we must develop and evaluate explanations of the origin of life from these clues, and continuously reevaluate as more clues are found.

Scientists today have proposed various explanations, of which we will discuss three. One explanation for the origin of life is that the first living cell, or cells, developed from nonliving matter according to chemical laws that we can observe today. This explanation is called the theory of **chemical evolution** or **prebiotic** (before biological life) evolution. The "chemical evolution" theory assumes that matter and energy somehow self-organized into complex forms without any outside intelligence directing the process. We call the process of self-organization without outside intelligence **spontaneous generation**. In most forms, the theory assumes that a very long time was needed to "test" millions of chemical combinations until the right combination for life was found. Since this theory is commonly taught in high school biology, we will study it in detail.

The second major explanation states that the first life on earth was designed by an outside agency, force, or intelligence. This theory suggests that life was formed according to an intelligent plan by an intelligent agent. Many who accept intelligent design as the best theory of life origins also believe that observations show natural processes are inadequate to account for the appearance of major types of living things, such as snails, clams, jellyfish, lampreys, fishes, amphibians, reptiles, birds, and mammals. Other design proponents agree, but don't rule out the use of the blueprints of previous organisms for these origins.

A third explanation is that life came to the earth from some other part of the universe.[2] Such life must have been relatively simple and able to travel in a spore-like form that could withstand the rigors of outer space. There are serious difficulties, however, with this proposal. Out of all of the diversity of life forms, few, if any, could withstand the radiation or extremes of heat and cold found in space for periods like those necessary for transport between solar systems.

In addition, the distances between stars and volumes of interstellar space involved are immense. The probability that spores released near one star would be intercepted by a planet orbiting another is not large. Furthermore, this proposal would explain only the appearance of life on earth; but not how life originated in the first place. The problem of life's origin would simply be shifted to another location.[3] For these reasons, we will not consider this last idea further.

In the remainder of this chapter we will study the two main explanations of origins—chemical evolution and intelligent design—beginning with an early form of spontaneous generation.

Spontaneous Generation—Science Does a "Double Take"

It seems far-fetched to us now, but for many hundreds of years people believed that all sorts of living things could originate suddenly, without parents, from mud or from decaying organic matter such as rotting meat. For example, decaying meat seemed always to be covered with swarming flies, so it appeared they originated from it. After the invention of the microscope, scientists could observe bacteria, and some thought that these, at least, originated spontaneously from nonliving chemicals in various broths believed to have been initially sterile—devoid of all life. The idea of spontaneous generation not only seemed to explain the appearance of new individual organisms, but was used by some to explain the origin of the first life on earth.

It is important to note that belief in spontaneous generation was based on direct, although incorrectly interpreted, observations and faulty experimental technique. Francesco Redi and Louis Pasteur demonstrated that certain claimed cases of spontaneous generation were not that at all. Redi's experiment in 1668 with maggots and rotting meat caused people to doubt the spontaneous generation of large (macroscopic) organisms. Two centuries later in the early 1860s, Pasteur's elegant experiments showed that the growth of bacteria and other microbes in "sterile" media was due to contamination of the flasks by air-borne, live, microbial (my-CROW-bee-al) spores. This meant the death of an idea that had been held, in one form or another, for thousands of years. Thereafter scientists rejected spontaneous generation as an explanation for the sudden appearance of such organisms.

Following Pasteur's experiments, our understanding of the complexity of living cells, including bacterial cells, greatly increased. Advances in cell biology and biochemistry during the rest of the 19th Century and the early years of the 20th provided additional reasons for ruling out the sudden origin of living matter. During this period, the view that life comes only from pre-existing life was universally accepted. Nearly every scientist agreed that cells came only from cells, and that even the simplest cell was not generated spontaneously. The idea of spontaneous generation was all but dead.

If the Darwinian theory of evolution is correct, then complex forms of life must have evolved by natural means from simpler ancestors. But how did the first life begin? Even if spontaneous generation does not occur now, did it happen at least once in the past, when conditions on the earth's surface and in its atmosphere were different? Some scientists never gave up on this possibility.

Over time, interest grew stronger in finding a natural explanation of life's origin. During the first two decades of this century, many advances were made in the study of viruses and the chemical nature of living matter (cytoplasm). Amino acids and other simple building blocks of cells were formed in the laboratory as sugars had been synthesized in the 19th Century. The important field of colloid chemistry was growing. Colloids are the particles of which gels are composed. The particle sizes of many colloids are approximately the same as the sizes of large cells. Some colloids were compared with cytoplasm and found to demonstrate similar properties. Through these and other studies, scientists gained more and more understanding of the chemical makeup of cells.

A Theory Is Born

On the basis of these advances, the Russian biochemist A. I. Oparin proposed a new form of spontaneous generation, a theory about the origin of the first life. In 1924 Oparin published his theory suggesting that the first cell or cells formed very gradually.[4] (When we hear the word "spontaneous," we may think of something sudden or that happens all at once, but the word simply means unaided by outside intelligence, regardless of the amount of time required.) As scientists' understanding of living cells became more sophisticated, it was recognized that the formation of something as complex as a cell must have required extended periods of time. According to Oparin, the atmosphere of the early earth was very different from the present one. Energy sources such as heat from volcanoes or lightning acted on simple carbon compounds in the atmosphere, transforming them into more complicated compounds in colloidal form. In the earth's early seas these newly formed compounds came together to form microscopic clumps, the forerunners of the first living cells on earth.

Essentially the same idea was put forward by the English biochemist J. B. S. Haldane in 1928.[5] He thought that ultraviolet light from the sun caused the transformation of simple gases (carbon dioxide, methane, water vapor, and ammonia) in the earth's primitive atmosphere into organic compounds, turning the primitive ocean into a hot, dilute "soup." (See Figure 1-1.)

Out of this soup came virus-like particles that eventually evolved into the first cells. Oparin and Haldane thus laid the foundation for the theory of chemical evolution which, despite many modifications in later years, is still called the *Oparin hypothesis* or the *Oparin-Haldane hypothesis*. Let's examine it in more detail. Note carefully the assumptions upon which it is built.

Oparin's Hypothesis

Assumption No. 1: Reducing Atmosphere

The earth's early atmosphere contained little or no oxygen.

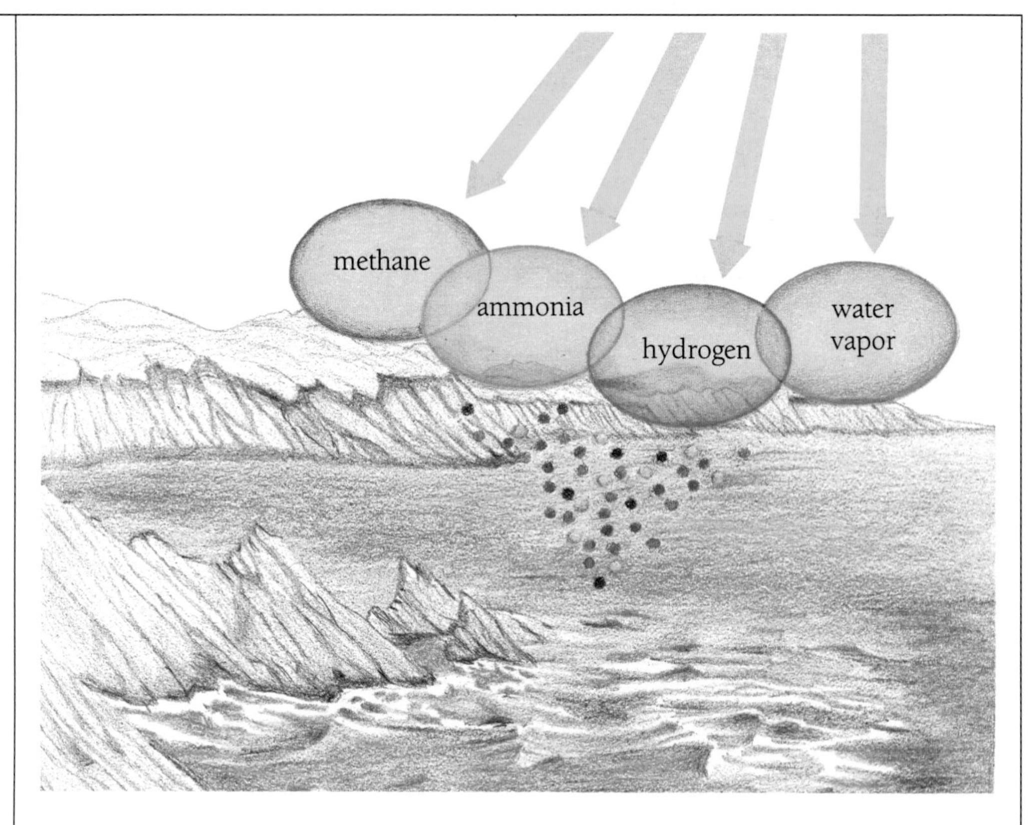

Figure 1-1. *As envisioned by Oparin's hypothesis, ultraviolet radiation was to have reacted with gases in the atmosphere to form important organic compounds on the early earth.*

According to Oparin, the first cells arose gradually over millions of years. He believed that conditions at the surface of the early earth allowed a massive accumulation of organic compounds before life began. Such an accumulation would improve the probability that all the right compounds could have come together and combined into a cell. However, it could not have occurred if the earth's atmosphere contained significant amounts of oxygen gas (O_2), because oxygen destroys organic compounds by reacting with them—a process called *oxidation*. Oparin believed that the early atmosphere must have been composed of gases such as methane (CH_4), ethane (C_2H_6), ammonia (NH_3), hydrogen (H_2), and water vapor (H_2O), but not oxygen. We call this kind of atmosphere a **reducing atmosphere**. He based his belief on the observation that hydrogen (H) is the most common element in the universe, and in all probability would have combined early in the history of the universe with other light elements to form the compounds listed above.

Furthermore, Oparin believed the first cells were *anaerobic* (able to survive without oxygen) and *heterotrophic* (unable to make their own food) obtaining many of their essential nutrients instead from the surrounding water. These anaerobic heterotrophs obtained their energy by *fermentation*, a method of releasing energy from organic molecules in the absence of oxygen gas.

Assumption No. 2: Preservation

The simple organic compounds formed in the soup were somehow preserved, so that the energy that caused them to form did not also destroy them.

The atmospheric gases could not have reacted together to form more complex compounds unless energy was available to cause them to react. Ultraviolet light from the sun, cosmic rays, electrical energy from lightning bolts, heat, and radioactivity might have provided the necessary energy. According to the theory, the available energy converted the atmospheric gases into more complicated compounds such as sugars, amino acids, and fatty acids. But the same energy sources also would have broken down the molecules of these compounds. The assumption is that these compounds were somehow protected from such destructive effects and collected in the earth's primitive oceans to form a soup out of which life could form.

Assumption No. 3: Reservation

Enough biological compounds were reserved for combination with the "right" molecules (rather than being tied up by reacting with useless molecules) to form the large molecules useful to life.

Oparin proposed that the simple organic compounds (biomonomers, bi-o-MOHN-uh-mers) accumulating in the primitive oceans combined with each other to form more complex substances such as proteins, nucleic acids (a very important one, unknown to him, is DNA), polysaccharides (long chains of sugar molecules) and lipids (fats). At first, these large biological molecules (biopolymers) were much simpler than their biochemical counterparts today, but gradually they became more and more complex. Eventually, there emerged proteins with **catalytic** properties—the ability to accelerate reaction rates without being harmed in the process. These were the forerunners of the first enzymes. It is assumed that amino acids reacted with each other to form larger molecules (polypeptides), and that simple sugars combined with other simple sugars to form complex sugars (polysaccharides). It is also assumed that DNA and RNA were formed in a similar way when their chemical building blocks combined.

Assumption No. 4: Uniform Orientation

Only "left-handed" or L-amino acids combined to produce the proteins of life, and only the "right-handed" or D-sugars reacted to produce polysaccharides, or nucleotides.

Sometimes quite early in their development, living cells showed a preference for a certain type of amino acid and a certain type of sugar. Two amino acids can be chemically identical, having the same chemical make up, but still differ in their three-dimensional shape, in the way the left hand differs from the right. In other words, they can be mirror images of one another. The protein which makes up living things *must* be composed of only left-handed amino acids (except glycine which is neither right- nor left-handed), even though both left- and right-handed forms are equally probable in nature. The same is also true of sugars; genetic material (DNA and RNA) can only include right-handed sugars.

Assumption No. 5: Simultaneous Origins

The genetic machinery that tells the cell how to produce protein and the protein required to build that genetic machinery both originated gradually and were present and functioning in the first reproducing protocells.

In currently living cells, both DNA and protein depend on each other for their existence. There are differing views as to how one could have originated without the other already being present. Some believe the genetic coding came first, others believe the functioning protein did, and still others maintain that both DNA and proteins appeared simultaneously.

Assumption No. 6: Specified Complexity

The highly organized arrangement of thousands of parts in the chemical machinery needed to accomplish specialized functions originated gradually in coacervates or other protocells.

Biological macromolecules (for example, proteins) combined to form complex microscopic particles called "coacervates" (co-A-ser-vates). Coacervates are organized droplets of proteins, carbohydrates, and other materials formed in a solution. Some scientists think that these particles were the forerunners of the first living cells. Coacervates have some of the properties of living things, and so may have "competed" with one another for dwindling supplies of "food" molecules in the primitive oceans. Oparin thought of this competition as a kind of Darwinian natural selection (see your biology text) which resulted in the survival and domination of ever more complex and life-like coacervates until a true cell finally appeared. These presumed first cells had cell membranes, complex metabolism, genetic coding, and the ability to reproduce. They finally dominated the primitive seas.

Assumption No. 7: Photosynthesis

A chemical system called photosynthesis, the process of capturing, storing, and using the energy of sunlight to make food, gradually developed within coacervates.

Speculation suggested that further development of some of the primitive *heterotrophic* (unable to make their own food from inorganic starting materials) organisms resulted in the formation of cells capable of *photosynthesis*. These were the first *autotrophic* (able to make their own food) organisms. According to the theory, the driving force for this evolutionary development was the gradual decrease of food molecules from the primitive oceans. It is assumed that spontaneous chemical events within the coacervates led to the formation of an efficient energy capture and processing system (photosynthesis). This system supplied the energy needs of the primitive cells. Since photosynthesis releases oxygen (O_2) into the air, it provided for the future development of heterotrophic organisms that used oxygen in respiration rather than having to depend on fermentation as a way of supplying energy.

In summary, Oparin's hypothesis visualized a very gradual origin of life from nonliving organic chemicals by a long process spread over hundreds of millions of years, and without the help of intelligent activity. Instead of cells appearing suddenly (as believed before Pasteur) the origin of life was viewed as a series of stages, each of which could be partially tested by experiments to assess its plausibility. Thus, Oparin's hypothesis was more than just a revival of the spontaneous generation idea; it was a refinement that attempted to explain the cell's complexity.

Future experimental results would assess the reasonableness of Oparin's hypothesis. While the hypothesis was very sketchy, Oparin himself, as well as the other adherents, were confident it would only be a matter of time before laboratory experiments would fill in the details of how life first developed.

What Experimental Studies of the Origin of Life Can Tell Us

As we have said, Oparin proposed that simple gases such as methane, ethane, ammonia, hydrogen, and water vapor, would react to form organic compounds when subjected to various energy sources likely to have been present on the earth before life began. This idea can be tested in the laboratory simply by enclosing these gases in a glass apparatus and bombarding the gas molecules with ultraviolet light or with electrical discharges, to see what compounds are formed in the apparatus. Such an experiment is called a *primitive atmosphere simulation experiment,* and many have been performed since the early 1950s.

The primitive atmosphere simulation experiments can only try to reproduce or simulate what we believe some of the conditions of the early earth were, and then show if the hypothesis is reasonable. Be sure to notice the term "simulation." These experiments can only try to simulate likely early earth conditions; they cannot provide observation of life's origin. Even though these experiments are called simulation experiments, people sometimes lose sight of this fact.

The Miller-Urey Experiments

Stanley Miller was a graduate student at the University of Chicago working under Professor Harold Urey (winner of the Nobel Prize in Chemistry in 1934). When Miller began his graduate work, nobody had yet carried out experiments to see if the primitive atmosphere assumed by Oparin would indeed produce organic compounds necessary for life. Miller and Urey were interested in doing just that. In order to duplicate the conditions which Oparin assumed to exist on the primitive earth, they designed the apparatus shown in Figure 1-2.

Water was placed in the bottom of the round-bottom, boiling flask. It provided water vapor to the atmosphere in the flask. Any trace of O_2 was eliminated. Methane and hydrogen gases were piped into the system, and ammonia gas was generated by heating dissolved ammonium hydroxide (NH_4OH) in the water at the bottom of the flask.

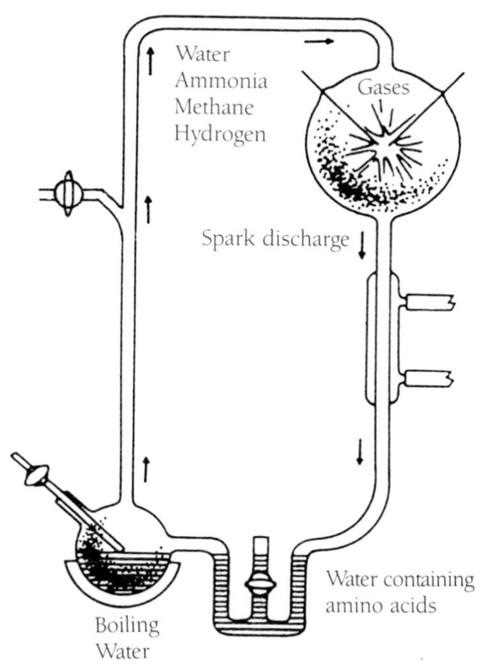

Figure 1-2. ***Miller's apparatus. The laboratory equipment used by Stanley Miller to perform his early prebiotic simulation experiments.***

During the experiment, the water in the flask was boiled, driving the gases in a clockwise direction through the apparatus. At the top of the apparatus was a 5-liter glass sphere containing two electrodes connected to a source of electricity. As the gases passed between these electrodes they were subjected to 50,000 volt sparks. Leaving the spark chamber the gases passed through a cooling condenser, which condensed water vapor and any nonvolatile (unable to form a gas) organic compounds that formed in the glass sphere. This solution then collected in a trap at the bottom of the apparatus. You can probably imagine the excitement as Miller waited to see the results of his first experiments. Among the resulting chemical products he identified in this solution were several of the amino acids found in proteins today. He identified glycine, alanine, and aspartic and glutamic acids. He also found several nonbiological amino acids, urea, and some simple organic acids such as formic, acetic, succinic, and lactic acids.

Since Miller's early work in the 1950s, many more biological compounds have been detected in primitive atmosphere simulation studies; the list now includes most of the basic kinds of organic compounds found in living cells.

One can readily see why the Miller-Urey experiments generated great excitement among scientists interested in the origin of life. These experiments seemed to prove that many of the chemical building blocks of life could have formed naturally under conditions assumed to have existed on the primitive earth (that is, a reducing atmosphere). Experimental evidence to support the first stage in Oparin's hypothesis was thus obtained, and the whole chemical evolutionary picture of the origin of life gained credibility and new adherents.

Problems with the Assumptions of the Oparin Hypothesis

But in spite of apparently promising results, many problems have since developed with the assumptions underlying the Miller-Urey experiments, some of them quite severe. Recall that Assumption No. 1 is that the earth's early atmosphere contained virtually no oxygen (O_2). If oxygen had been present in the earth's early atmosphere (even 1% by volume compared to 21% today), it would have been impossible for organic compounds to have accumulated the way they did in Miller's experiment. Such compounds would not have formed at all or would have been quickly destroyed by oxidation. Strong chemical arguments have been set forth favoring the presence in the early atmosphere of oxygen in significant amounts.[6] Recently discovered geological evidence indicates that significant amounts of oxygen may well have been present in the earth's atmosphere at the same time that the first life was supposed to be developing. While the geological evidence for oxygen in the early atmosphere is not conclusive, it is not at all clear that Oparin's basic assumption of an early oxygen-free atmosphere is correct. Unfortunately, high school biology textbooks fail to mention the direction that recent studies of the early atmosphere have taken.

Moreover, had oxygen been present in Miller's apparatus during his experiments, he could not have produced the organic building blocks of life; no biological compounds would have accumulated in the apparatus, and had enough oxygen been present with the hydrogen gas, a spark would have caused an explosion.

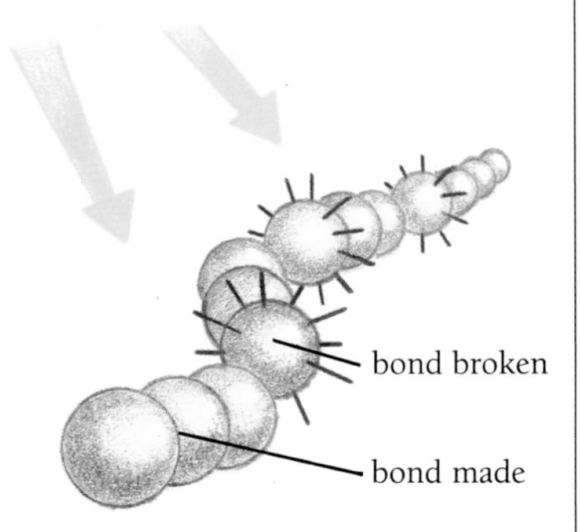

Figure 1-3. *The same energy that would cause amino acids to react in a primordial soup, forming short polypeptides, would soon break down those developing polypeptides with a little additional exposure.*

A second problem with the Miller-Urey experiment was that the design of the experimental apparatus didn't really match what has been understood of the prebiotic conditions. Amino acids and other simple nonvolatile, stable organic compounds accumulated in the trap, where they were preserved from the destructive effects of the electrical discharges. If the amino acids and other products had been continuously exposed to the energy source, as would their counterparts in the early earth, they would have been destroyed as they were being formed, and Miller could not have detected them in the solution he produced in his experiment (see Figure 1-3).

In this respect, the design of Miller's apparatus was very different from the conditions existing on the earth as proposed by Oparin.[7]

It's not a matter of one organic molecule in the early earth somehow surviving the destructive effects of energy. The key issue is one of *equilibrium* or balance between opposing results. Energy is like a two-edged sword which cuts both ways. On the one hand, it builds up complex molecules out of simpler parts; but on the other, that same energy breaks up developing molecules. The end result is a balance between the constructive and destructive effects, keeping most molecules relatively simple. It is much like the balance in a checking account. You have to take both deposits and withdrawals into account; you cannot write checks without regard to how much you have on deposit. But for many years, textbook expectations about how complex the molecules in the chemical soup would become neglected the destructive effects of energy. Most of today's books don't even mention the destructive effects. Actually, when we take into account these destructive effects, we recognize that the equilibrium state of the soup would not favor complex molecules, but simple ones, of little use in spontaneously forming the machinery of a living cell.

The major energy source for the presumed synthesis of organic compounds on the primitive earth is believed to have been ultraviolet (UV) radiation from the sun. But if the proposed methane gas were actually present in the early atmosphere, some of the ultraviolet radiation entering the earth's atmosphere would have been absorbed by this gas many miles above the earth's surface. Any products (for example, amino acids) formed from methane would have appeared high in the atmosphere, far from the oceans below. If these products were to be saved from destruction by the ultraviolet radiation bathing the atmosphere, they would have had to be dissolved rapidly in the oceans. Because of their altitude at the time of formation, and the convection currents in the atmosphere, the vast majority of newly formed products would take long periods to reach the oceans below. Meanwhile, the methane products would have been exposed to the destructive effects of the incoming UV radiation, and may never have reached the oceans at all (see Figure 1-4).

Even if small amounts of organic compounds had reached the ocean, by no means would they have been safe there, for at least two reasons: 1) some ultraviolet radiation would have penetrated beneath the ocean surface, destroying some of the dissolved compounds; 2) organic compounds, such as amino acids, tend to break down when dissolved in water. The higher the temperature, the faster this breakdown occurs.

We can see that Miller's experimental design was faulty. The trap used in his apparatus did not realistically correspond with any reasonable protective mechanism presumed to have existed on the early earth.

A third problem with the Miller-Urey experiment involves the third assumption of the Oparin hypothesis: "Reservation. Sufficient quantities of biological compounds were reserved for combination with like molecules, or the 'right' molecules for the formation of the large molecules useful to life." How helpful it would be to the theory if molecules could be put on "lay-away." But reactions in an organic soup full of tens of thousands of varieties of chemicals, instead of neatly matching amino acid to amino acid and sugar to sugar, would tie up useful chemicals in cumbersome masses of cross-reactions.[8]

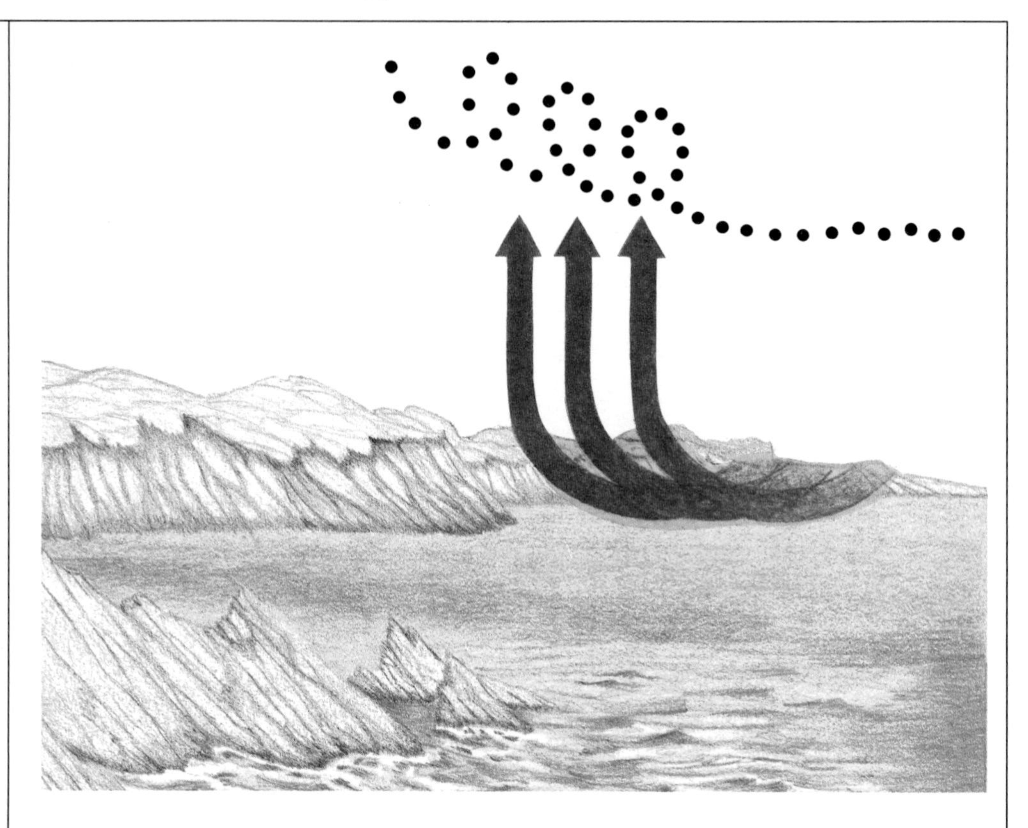

Figure 1-4. ***Convection currents would prolong the exposure of organic compounds to ultraviolet radiation.***

In keeping with this expectation, no biopolymers (useful to life) have been found in the Miller experiments to date, except some very small peptides. (They have been produced, however, in specially modified experiments such as those discussed below). But in the Miller experiments, large yields of nonbiological, amber-colored goo are produced. This material has not been identified, but it quite possibly results from the cross-reaction of aldehydes and cyanides (both known to be formed in the Miller apparatus, and also known to form the amber macromolecule *aldocyanoin*), or from amino acids and sugars (which combine to form the nonbiological substance *melanoidin*). Therefore, interfering cross-reactions do occur under experimental conditions, casting real doubt on Assumption No. 3.

It is worth noting also that we have no geological evidence of any massive pre-life (prebiotic) accumulation of organic matter. The clay deposits of the time, found in abundance, would have retained large amounts of hydrocarbons and nitrogen-rich compounds from the prebiotic soup. The surface of the clay has tiny cavities that would have imprisoned these molecules where they would still be evident today. Thus if the "prebiotic soup" had really existed, we would expect to find such surviving traces of it in the oldest rocks, but we do not.[9] Few, if any, biology textbooks inform students of this fact.

A fourth problem with Miller's experiment involves Assumption No. 4, Uniform Orientation. This assumption states: "Only 'left-handed' amino acids (L-amino acids) combined to produce the proteins of life." The amino acids that make up the proteins of living cells are all "left-handed" (the L-form), but in the Miller-Urey experiments, the amino acid products found in the apparatus are always a racemic (rah-CEE-mick) mix-

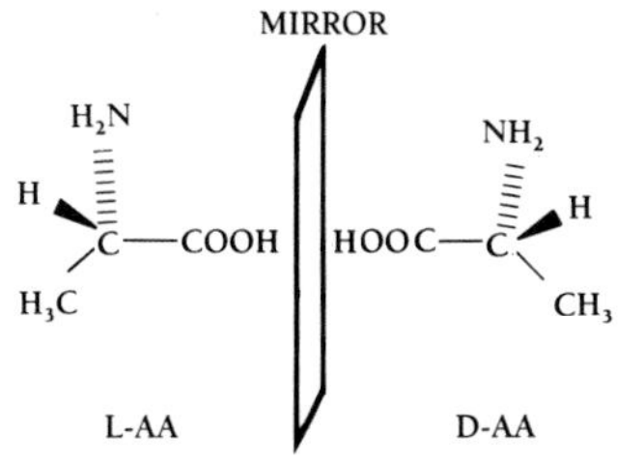

MIRROR

H_2N, H, H_3C — C—COOH | HOOC—C — NH_2, H, CH_3

L-AA | D-AA

Figure 1-5. *The L- and D- forms of alanine. While the proteins of living things are composed of only the L-forms, both forms are found in prebiotic simulation experiments.*

ture, that is, 50% "left-handed" and 50% "right-handed" forms. The L- and D-forms of the amino acid alanine are shown in Figure 1-5. No one knows why amino acids in living things only occur in one of their two possible forms.

In other kinds of primitive earth simulations, sugars have been produced in similar racemic mixtures. Living matter includes only "right-handed" (D-sugars), but the sugars found in the experiments are always mixtures of the two (L- and D-) forms. How did living things come to need only one form? Scientists have conducted special experiments to try to find out how these "preferences" originally developed, but so far all such attempts have failed.[10]

Regardless of the numerous problems with the Miller-Urey experiments, many scientists agree that the appearance of amino acids in them gives experimental support to their belief that life began in some spontaneous way.

Proteinoid Microspheres

The next problem faced by the Oparin hypothesis relates to Assumption No. 5, Simultaneous Origins. How can we explain amino acids joining together to form the first proteins? The great problem here is that amino acids will join together, but not in the right sequences without the help of enzymes that act as specific catalysts. But these enzymes themselves are very special proteins that would have required DNA to code for their catalytic ability. How then did the first amino acids form enzymes when there were neither existing enzymes to serve as catalysts in linking the protein structure together, nor DNA to code for the structure? (See Figure 1-6.) Simply put, this is a classic "Which came first, the chicken or the egg?" problem. To get around the problem, some have suggested that RNA came first, since, besides its code-carrying capacity, it also exhibits some catalytic properties![11] So perhaps RNA came first, carrying the code for proteins and also assisting in their synthesis. After some initial enthusiasm for this suggestion, many leading scientists, including former defenders of the idea,

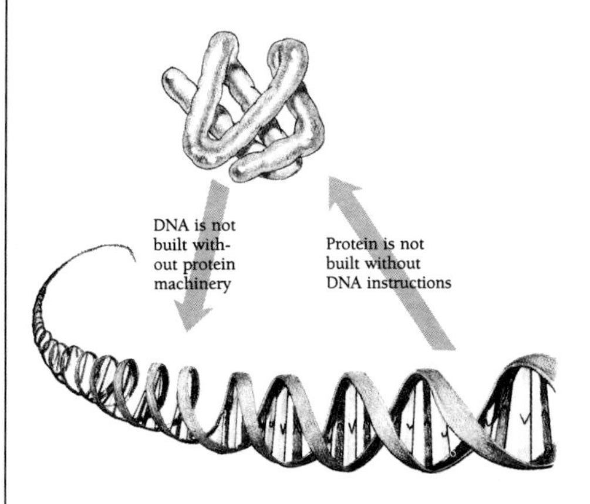

Figure 1-6. *Both proteins and DNA are absolutely dependent upon the prior existence of the other for their own existence.*

have discarded it, citing the same objections given above to the notion that DNA came first![12]

To solve this problem, scientists have tried to devise ways to link amino acids together into protein-like molecules without using either existing enzymes or DNA. One such experiment was performed by Sidney W. Fox while he was at the University of Miami.

Dr. Fox heated dry mixtures of 16 to

18 amino acids at 160-180° C for several hours in a nitrogen atmosphere. He only used the kind of amino acids that form proteins in living things. He found that they joined together by the loss of water molecules to form polypeptides (long chains of amino acids). He called these polypeptides "**proteinoids**," since they faintly exhibit some of the properties of true proteins. For example, the proteinoids made some kinds of chemical reactions go a little faster than they would otherwise. Fox believes these proteinoids represent the beginnings of enzyme activity on the primitive earth.

If proteinoids are dissolved in boiling water, and the solution cooled, the proteinoid molecules will aggregate (clump) together to form uniform microscopic spheres about the size of bacterial cells (see Figure 1-7). Fox believes these little spheres represent the first step toward cellular life on earth.

The process of forming proteinoid microspheres from amino acids (assuming for the moment that amino acids could have accumulated on the primitive earth) is thought to have taken place near volcanoes. It is suggested that the amino acids might have formed when the gases of the primitive atmosphere came in contact with molten rock in a volcano (at a temperature of about 1200° C). The amino acids then accumulated a few miles away from the center of the active volcano where surface temperatures in the range of 160-180° C might occur. At this temperature the dry amino acids condensed into proteinoids which were then protected from the destructive effects of heat by the cooling action of rainfall. Once suspended in pools of water, the proteinoids would form microspheres that would compete with one another. Those fortunate enough to develop life-like qualities would survive in the long run. The process could occupy millions of years, although Fox has suggested it could take but hours to get to the protocell stage. Eventually, the first self-reproducing, fer-

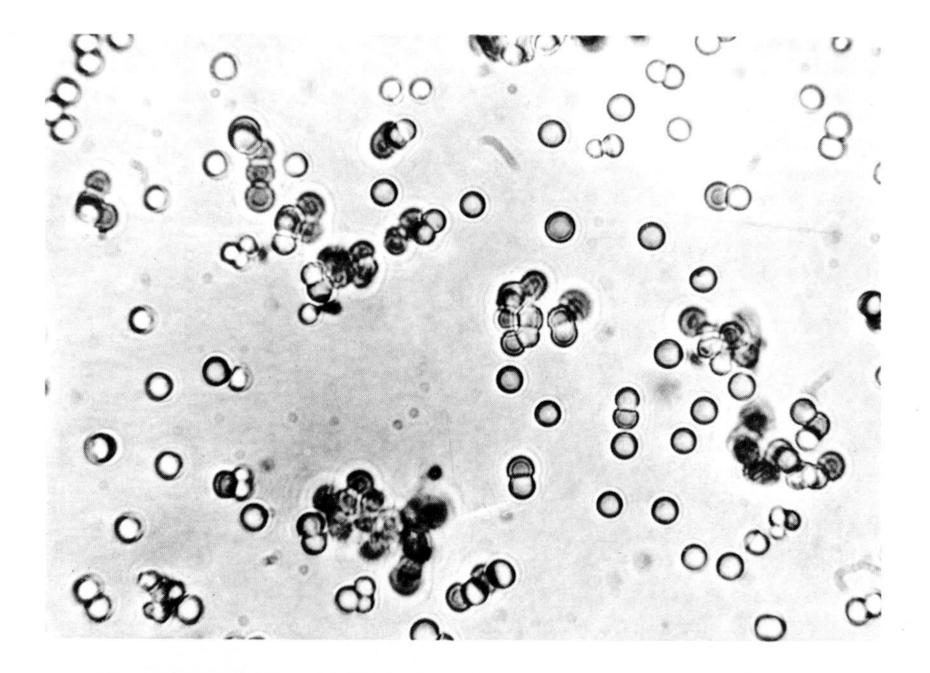

Figure 1-7. *Optical micrograph of proteinoid microspheres.*

menting cells would be formed, and life itself would have gained its first foothold.

In theory, Fox's experiments cover the transformation from the amino acid stage demonstrated by the Miller-Urey experiments all the way to the threshold of life. Many scientists believe that the results of these two types of experiments provide strong support for Oparin's hypothesis. But other scientists disagree. Let us look at some of the reasons for the disagreement.

Problems with Proteinoid Microspheres

First, in his experiment Fox used mixtures containing only *protein-forming L-amino acids*. Where on the primitive earth could such a mixture have occurred? We have already seen that thousands of interfering cross-reactions would have occurred in the soup, preventing the successful fulfillment of Assumption No. 3 of Oparin's hypothesis. These would have tied up protein-forming amino acids like Fox used. Some of these substances would have combined directly with amino acids, thus blocking the formation of proteinoids. For example, we have already seen how sugars react with amino acids to form the nonbiological compound known as melanoidin. Because of such cross-reactions, which Oparin assumed *did not occur*, it is very unlikely that proteinoids could have formed under natural conditions on a primitive earth. In short, many scientists agree that Fox's use of selected and purified amino acid mixtures isn't realistic.[13]

Second, there is disagreement about the proposed sequence of events which supposedly occurred near volcanoes. The required combination of high and low temperatures, with rainstorms occurring just at the right time and place, seems unrealistically "choreographed" and highly improbable to many scientists. Even if proteinoids did form, the heat which formed them would also have destroyed them or they would have broken down spontaneously before they could have played a role in the formation of life.

Most important of all, even if these problems were solved, big differences exist between proteinoid microspheres and the very simplest living cells.[14]

Assumption No. 6 said, "The highly organized arrangement of thousands of parts in the chemical machinery needed to accomplish specialized functions originated gradually in coacervates or other protocells." But how did it originate? True cells contain 500 or more different enzymes, each specialized to carry out a single chemical reaction with great speed and efficiency. All of the enzymes work together like the parts of a well-designed machine in carrying out the cell's metabolic activities. Each enzyme consists of a precise sequence of L-amino acids. The sequence is important because it gives the enzyme its character and determines its function, just as the combination to a given lock must always have the same sequence of numbers to work. The instructions for building specific enzymes with the correct sequence are carred in the DNA molecules ("genes") of the cell. DNA is built from subunits, the nucleotide bases that make up, three at a time, genetic "words." The "words" specify the kind and order of amino acids to be found in all the proteins of the cell. Thus it takes 300 "letters" to specify an enzyme of 100 amino acids—a smaller than average enzyme.

What's more, in order to build a single protein or DNA molecule in a cell, about 60 specific proteins acting as enzymes are needed. If even one is missing, proteins will not be formed. In other words, all of the required proteins must already be present in a cell before other proteins can be formed. The remarkable process by which proteins are synthesized is diagrammed in Figure 1-8.

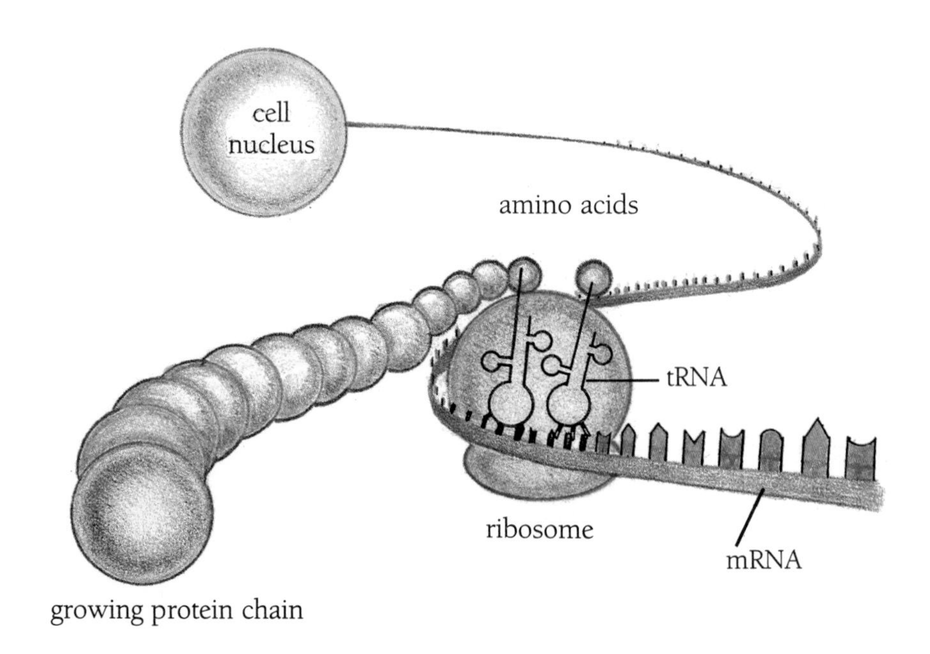

Figure 1-8. ***Protein synthesis. A view from inside the cell of the assembly of a protein molecule from the DNA pattern brought by mRNA to the site of assembly.***

Where did all the many proteins needed to build other proteins first come from? The answer (according to the Oparin hypothesis) is that they were assembled spontaneously. But the probability of forming just one specific protein in an undirected search is practically zero.

Consider a small protein consisting of 100 amino acid units. How many different combinations of the basic 20 amino acids are possible in a chain of 100 units? The answer is $20^{100}=10^{130}$ (1 followed by 130 zeros). This number is so enormous that there has not been enough time during the conventionally accepted age of the universe (15 billion years) to try every combination in an attempt to find the specific combination of one protein!

It is recognized, however, that some amino acid variation is possible at some positions in the chain without disrupting the protein's function. Nonetheless, scientists have now both calculated and demonstrated empirically that this tolerance would increase the probability of forming a 100-unit, functional protein only to one in 10^{65}.[15] This probability is still so minuscule that there would not be enough time to make even one such protein by chance. You can begin to appreciate how big a problem this is when you realize that, not just one, but about 60 specific proteins are needed *before* the precise division of labor of a cell's protein-making machinery (including translation and transcription) can successfully function! And, not only must those roughly 60 specific proteins all exist at the same time, they must also occur together in the same tiny region of the cell.

Oparin's hypothesis assumes that all of the necessary enzymatic proteins were formed by some type of natural process in spite of the overwhelming odds against each such occurrence. The laboratory work of Fox and others is doubtful support for the idea that spontaneous chemical processes could have formed the kinds of proteins needed to run the DNA and protein-making machinery in a cell.

Cell Membranes

True cells are surrounded by a complex cell membrane. These membranes are made of special fat molecules called *phospholipids* (foss-fo-LIP-ids) together with many classes of specialized protein molecules. Unless the protein molecules are of specific classes, they will not function properly. By contrast, proteinoid microspheres have an outside boundary which is much thicker and is made only of proteinoids, very simple molecules compared to true proteins.

As you can see, there is serious difficulty with the idea that living cells evolved by natural means from a primitive organic soup. There is neither solid theory nor promising experimental basis for the belief that specified complexity—the highly organized arrangement of thousands of parts—arose spontaneously.[16] We can safely conclude that Oparin was most probably incorrect in making Assumption No. 6, underlying the chemical evolution of such cellular machinery.

The final assumption of Oparin's hypothesis is also still without support. Assumption No. 7 states that some cells eventually develop the ability to capture sunlight and transform it into food (photosynthesis). There is no experimental evidence to support the view that such a complex process developed by presently known natural means. In fact, the step-by-step formation of such a complex energy-processing system is highly improbable, even over billions of years.

Scientific Case for the Intelligent Design of Life

There is a reasonable alternative explanation for life's origin, an explanation that has scientific support. As we observe how living things function, we are impressed by the high levels of complexity and organization that are necessary. The organization in a living creature is an expression of the information carried in the genetic material of a cell as it directs the building of its parts. This process is similar to the building of a house according to an architect's plan, or the writing of a book. In 1967, scientist Michael Polanyi said:

> A book or any other object bearing a pattern that communicates information is essentially irreducible to physics and chemistry... We must refuse to regard the pattern by which the DNA spreads information as part of its chemical properties.[17]

Today this understanding, now shared by many scientists, can be formalized in the following broad generalization about the behavior of matter and energy: "Information never arises from physical or chemical causes alone." This is the same kind of formalized description of the behavior of matter and energy as are the laws of thermodynamics—a broad generalization about what matter and energy will or (in this case) won't do. Moreover, it is subject to test in an identical manner. The second law of thermodynamics could be overturned by the repeated observation of exceptions to the rule. Likewise, one repeatedly confirmed observation of physical or chemical processes giving rise to information would disprove this generalization.

Presently, however, there are many good reasons to believe this generalization. First, we know of no exceptions to it. Second, accepted concepts from information theory seem to strengthen this conclusion. We know that energy flowing through some systems may produce highly ordered patterns. Darwinists have pointed to these patterns as suggestive of how spontaneous generation may have occurred. But informational sequences are different. They manifest an *irregular order*, reflecting the constraints of a coded message. Confusing the two ideas, order and information, has led many to attribute properties to brute matter that it does not possess. As information scientist Hubert Yockey cautions:

> Attempts to relate the idea of order . . . with biological organization or specificity must be regarded as a play on words which cannot stand careful scrutiny.[18]

One hundred and fifty years ago, nearly everyone acknowledged true design in the universe. But Charles Darwin's argument from natural selection said that there was only "apparent" design in living organisms. It may be, however, that the discussion will soon come full circle.

What experimental evidence exists in support of the view of intelligent design? First of all, the experimental tests of the Oparin hypothesis produced results which more reasonably support the intelligent design view. For example, when the scientific investigator doesn't interfere with the experiment, the experimental production of amino acids results in 50:50 mixtures of the right and left-handed forms. Intelligent restrictions on the experiment, however, can yield primarily one form and not the other.[19] These results support the idea that the exclusive production of left-handed forms resulted from intelligence. In fact, apart from intelligence using selected chemicals and controlling conditions, amino acids have not been collected in the laboratory. There is no reason to doubt that this was true, too, at the time of life's origin.

An Analogy

The case for the intelligent design of life can be clarified by an analogy. Suppose we left a brand new pickup truck in a clearing near a native village too remote for any contact with civilization, and suppose we left the keys in the ignition and the tank full of gas. Soon a villager or two would happen upon it. They run to the village and bring back their fellow tribespeople. Together they explore the truck, inside and out for hours. Losing all track of time, they tinker with knobs, stare into mirrors, push buttons, climb in and out, under and over the truck, talking excitedly all the while. Finally night comes, and the fascinated aborigines return to the village. But they will be back. In fact, it's not hard to imagine that they begin to visit the truck repeatedly for years. Within the first week and quite by accident, they learn to turn the ignition key, start, and drive the truck. But even more impressive, they eventually learn to take the truck apart and put it back together, coming to understand it mechanically. None of this is beyond reason; they could come to a nearly complete understanding of the mechanics of the truck. *But this would be something quite other than knowing where it came from in the first place.*

If the villagers tried to determine the truck's origin, it would come as no surprise if they wondered *who* was responsible—if they envisioned some kind of intelligence like their own. In so doing, they would actually be following sound principles of investigation, the principles of *cause and effect* and of **uniformity** (analogy). The first principle states that every effect (in this case, the truck) has an adequate cause. The second principle, uniformity, states that the cause of a given effect today is uniform (analogous) with the causes of similar effects in the past. If effects from the past and present are the same kind, and the cause of a presently observed effect is known, then it is considered reasonable (the philosopher David Hume called it proof[20]) to assign the same kind of cause as responsible for the effects have like causes.

Though perhaps unimpressive by computer-age standards, even a primitive tribe has its own technology, such as bows and arrows, huts, drums, levers and fulcrums, small animals traps, and many other implements. It would be perfectly rational and most likely to occur to these villagers that, just as it took their own intelligence to make these implements, it must have taken some similar intelligence to make the truck. While we wouldn't expect them to be able to state the principles above, they would nevertheless be employing them if they inferred an intelligent maker for the truck.

What does all this say about how life originated? For a moment, consider how living organisms and manufactured products both exhibit the property of organization or specified complexity. The pickup truck has many parts that make up a working whole and they all obey discoverable physical and chemical laws; *but the truck does not form spontaneously as a result of these laws*. For example, an internal combustion engine does not come into existence because of the physics of the metals from which it is made. Instead, an engineer or team of engineers had to take the physical properties of these metals into account in arriving at the design and manufacture of the engine. Is the kind of complexity found in living cells more nearly like the complexity of proteinoid microspheres or pickup trucks? Pickup trucks, of course.

The chemical properties of amino acids, sugars, purines, pyrimidines, etc. are not the properties needed to form genetic coding and the protein-synthesizing machinery. In fact, the natural chemical tendencies of these compounds as they would have been expressed in a hypothetical "primitive soup" would have *inhibited* the formation of coded messages. For example, amino acids would have reacted with sugars, preventing the formation of DNA and RNA.

We do find numerous examples of order (repeating patterns, symmetry, etc.) in the chemistry that would be found in a primitive soup and elsewhere in nature. However, nowhere in nonbiological nature do we find specific complexity that is even roughly analogous to coded information. In fact, the only comparable analogy known for the genetic code does not occur naturally. It has been discovered that the structure of information in living systems is mathematically identical to that of written language.[21] Since both written language and DNA have that telltale property of information carried along by specific sequences of 'words,' and since intelligence is known to produce written language, is it not reasonable to identify the cause of the DNA's information as an intelligence too?

When the available evidence is taken into account, it seems highly probable that the origin of life on earth involved the fashioning of molecular complexity in a way similar to the production of manufactured items. In fact, the living cell (even the very simplest one) has the complexity of a miniaturized, automated factory. We should no more expect the spontaneous emergence of molecular "machines" such as the DNA, aminoacylsynthetases (a-ME-no-A-sil-SYN-thuh-ta-suhs), transfer RNAs, ribosomes, etc. from simple organic compounds than the spontaneous assembly of robotic tools in an automated automobile factory from raw rubber, steel, silicon, etc. In both cases, a complex set of engineering designs *is required*, designs that were intelligently created. Then further intelligent activity is required in the assembly process.

Well-designed experiments on the origin of life should continue. Modern ideas of spontaneous generation or chemical evolution, however, do not realistically account for the appearance of biological complexity in prelife chemical systems. Thus it was no surprise when, in contrast to the optimism of textbook writers, one of the leading experimental researchers in origin of life studies, Klaus Dose, summarized his review article on this field as follows:

> More than 30 years of experimentation on the origin of life in the fields

of chemical and molecular evolution have led to a better perception of the immensity of the problem of the origin of life on Earth rather than to its solution. At present all discussions on principle theories and experiments in the field either end in a stalemate or in a confession of ignorance.[22]

As we have seen, many of the most important assumptions underlying the idea that life originated by nonintelligent processes do not correspond to the facts of science, and are not supported by sound reasoning from those facts. Some scientists protest such statements, maintaining that in the future discoveries will be made that will essentially circumvent present findings. This idea has been called "promissory materialism." And while no one can say for sure that this won't happen, science cannot confidently proceed by discounting what is known in favor of hoped-for future discoveries.

On the other hand, the experimental work on the origin of life and the molecular biology of living cells is consistent with the hypothesis of intelligent design. What makes this interpretation so compelling is the amazing correlation between the structure of informational molecules (DNA, protein) and our universal experience that such sequences are the result of intelligent causes. This parallel strongly suggests that life itself owes its origin to a master intellect.

Suggested Reading/Resources

The Mystery of Life's Origin: Reassessing Current Theories, by Charles B. Thaxton, Walter L. Bradley, and Roger L. Olsen. New York: Philosophical Library Publishers, 1984 and Dallas: Lewis and Stanley. A scholarly critique of chemical evolution.

Origins: A Skeptic's Guide to the Creation of Life on Earth, by Robert Shapiro. New York: Summit Books, 1986 (hardbound) Bantam Books, 1987 (paperbound). Another good critique of chemical evolution.

References

1. H. Rolston, 1988. *Zygon: Journal of Religion and Science* 23,352.
2. S. Arrhenius, 1908. *Worlds in the Making*. New York: Harper and Row, and F. H. C. Crick and L. E. Orgel, 1973. *Icarus* 19,341.
3. C. Thaxton, W. Bradley and R. Olsen, 1984. *The Mystery of Life's Origin*. Dallas: Lewis and Stanley. pp. 191-196.
4. A. I. Oparin, 1924. *The Origin of Life* (Russian Proiskhozdenic Zhizny), Moskovski Rabochii, Moscow. The best general Oparin reference is, A. I. Oparin, 1957. *The Origin of Life on Earth*, 3rd ed., New York: Academic Press.
5. J. B. S. Haldane, 1928. *Rationalist Annual* 148,3-10.
6. H. Clemmey and N. Badham, 1982. *Geol.* 10,141-146, and J. H. Carver, 1981. *Nature* 292,136-138.
7. Thaxton, Bradley and Olsen, p. 102.
8. Ibid., pp. 104-106.
9. J. Brooks and G. Shaw, 1973. *Origin and Development of Living Systems*. London and New York: Academic Press, p. 359.
10. W. Bonner, 1991. *Origins of Life* 21,59-111.
11. T. R. Cech, 1986. *Proc. Natl. Acad. Sci. U.S.A.* 83,4360, and T. R. Cech and B. L. Bass, 1986. *Ann. Rev. Biochem.* 55,599.
12. R. Shapiro, 1988. *Origins of Life and Evolution of the Bioshere* 18,71-85, and M. Waldrop, 1989. *Science* 246. 1248-1249.
13. Thaxton, Bradley and Olsen, p. 162.
14. Ibid., pp. 174-176.
15. J. U. Bowie, R. T. Sauer, 1989. *Proc. Natl. Acad. Sci. U.S.A.* 86,2152-2156, et al., 1990 *Science* 247,1306-1310, and J. F. Reidhaar-Olson, R. T. Sauer, 1990. *Proteins: Structure, Function, and Genetics* 7,306-316.
16. K. Dose, 1988. *Interdiscipl. Sci. Rev.* 13,348-356.
17. M. Polanyi, 1967. *Chem. & Eng. News*, Aug. 21, p. 62.
18. H. Yockey, 1977. *J. Theoret. Biol.* 67,377-398.
19. Thaxton, Bradley and Olsen, pp. 156-162.
20. D. Hume, 1748. *Inquiry*, ed. R. M. Hutchins, *Great Books of the Western World*, Vol. 35. [1952] Chicago, p. 458.
21. H. Yockey, 1981. *J. Theoret. Biol.* 91,13-16.
22. K. Dose, 1988. 13,348.

EXCURSION CHAPTER 2

Genetics and Macroevolution

Introduction

What kind of children will result from the marriage of a tall and a short person? Common wisdom often predicts that the children from such a marriage will be medium in height because of "mixed blood." In other words, some people still think of heredity as a mixing or blending of parental traits.

Darwin's View of Inheritance

Charles Darwin held that view. He believed that hereditary variations disappear permanently when opposite types mate. According to his theory of blending, offspring should be intermediate between their parents not only in physical appearance, but also in whatever hereditary material they receive and pass on. For example, according to the theory of blending, what should happen if a red morning glory is crossed with a white one? The offspring should be blended, and therefore pink in color, as indeed some would be. Now what should happen if these pink plants are crossed with each other? According to Darwin's thinking, the offspring from this cross should also be pink, and so on through the generations, the red and white color having been lost forever. Happily, such questions are a part of empirical science, so we can test the blending or mixing theory. In actual practice, the red and white colors do reappear in the second generation, proving that the theory of blending is false.

Darwin's book, *The Origin of Species*, did more to establish evolution as a scientific theory than any other single scientific work. No study of modern biology can reasonably overlook Darwin's thought. In *The Origin of Species*, he strongly argued that no existing species has been individually created. Instead, said Darwin, every species has descended from some preexisting species by the process of **natural selection**, and all species can trace their ancestry back to several original "beings." According to this theory, new traits in organisms arise when natural selection picks and and chooses from numerous natural genetic variants. Certain ones, the theory says, will be passed on to future generations if they give their possessor a competitive edge or advantage over other organisms of the same general type, and leave more offspring. Darwin assumed that, as this process proceeded, advantageous new traits would accumulate until a new species is formed. It is important to note, however, that this was only a logical conclusion or deduction, and was not established by empirical observations, despite the title of his book.

Not until the 6th edition of his book did Darwin give evidence he had become aware that his theory of natural selection was incompatible with this view of heredity. According to natural selection, an advantageous new trait would have to be preserved and passed on undiluted to future generations. For example, during the Industrial Revolution in England, soot from the factories covered the trunks of nearby trees. The peppered moth (*Biston betularia)* was now more easily spotted and eaten by birds because the moths were light and the tree trunks upon which they would rest had become sooty black. If a dark variant moth were to appear it would have a **selective advantage** because it could not be as readily seen against the black tree trunks. However, if the blending theory were true, the advantageous dark trait would be lost in following generations when the dark moth mated with a light moth; blending would cause that trait to be lost forever. The first generation would be gray, and future generations would be gray or lighter, depending on the color (gray or white) of the first generation's mates. Darwin's acceptance of the idea of blending inheritance actually made evoluton by natural selection impossible, although for some years he failed to realize it.

Mendel's Discoveries

Living at the same time as Darwin, an Austrian monk named Gregor Mendel spent a great deal of time crossing different varieties of garden peas and analyzing the results of the crosses. Mendel conducted his experiments with great care and based his conclusions on scientific evidence. He brilliantly concluded that inheritance in the garden peas is determined by six principles:

1. The inheritance of traits is determined by (what were later termed) genes that act more like individual physical particles than like fluid.
2. Genes come in pairs for each trait, and the genes of a pair may be alike or different.
3. When genes controlling a particular trait are different, the effect of one is observed (dominant) in the offspring, while the other one remains hidden (recessive).
4. In gametes (eggs and sperm) only one gene of each pair is present. At fertilization gametes unite randomly, which results in a predictable ratio of traits among offspring.
5. The genes controlling a particular trait are separated during gameteformation; each gamete carries only one gene of each pair.
6. When two pairs of traits are studied in the same cross, they are found to sort independently of each other.

While Mendel's principles have been expanded and redefined, they still remain basically sound today. It is unfortunate for Darwin that he did not learn about Mendel's work sooner. And it is especially ironic since he had a copy of Mendel's papers in his library, which he never read. (Recent examination showed that its joined pages were left uncut.) We might wonder whether Darwin would have discarded his own theory of blending in favor of Mendel's view of heredity, had he studied Mendel's principles. Moreover, he might have reexamined his own theory of natural selection and gone along with Edward Blyth's (see section on Natural Selection and the Fixity of Species, p. 67) original view that natural selection was a conservative force. He would have discovered a means by which new traits could be preserved and passed on in successive generations. He would also have discovered a genetic system more bent on preserving existing traits than on generating new ones.

Before taking a closer look at the genetic world, it will help us to recognize that there are two levels or categories of evolution, **microevolution** (small-scale evolution) and **macroevolution** (large-scale evolution). The shift in populations of moths from light coloring to dark illustrates microevolution: an increase in the frequency of a trait based on a relatively small shift in the gene frequency of a population. Moreover, it is quite likely the **gene pool** of the moth population already contained the genes for the darker coloring. If so, the "new" darker coloring involved no novelty (no truly new trait) and required no new genetic information. In contrast, macroevolution refers to wholesale changes of physical and behavioral characteristics requiring the input of new information and/or greater complexity. Darwinists see the means for this advance in complexity as a natural, undirected process, whether through few or many intermediate steps. The occurence of microevolution is little debated between Darwinists and intelligent design proponents; it can be observed, and nearly every scientist of either view acknowledges it. What is at issue then, is macroevolution, and our task in this chapter is to compare this concept with the operation of established genetic principles discovered by Mendel and others.

The Discovery of DNA

Mendel laid the groundwork upon which our modern understanding of genetics has been built. In 1953, two scientists, Francis Crick and James Watson, discovered the architecture of the now famous double helix DNA molecule. Although they knew it was somehow responsible for the diversity of living organisms, it was not yet clear how it could determine the sequence of amino acids in proteins, which was the key to the structure of living things. They did show, however, the composition and architecture of DNA molecules. DNA molecules are composed of two each, of two kinds of nucleotide bases: purines, quanine (G) and adenine (A) and pyrimidines, thymine (T) and cytosine (C) (see Figure 2-1). When combined in molecular form, they become two chain-like strands running in opposite directions and spiraling around each other. The two strands join at regular intervals, by complementary bases, one from each strand. The bases come together or "pair" by weak hydrogen bonds. It was later worked out that three

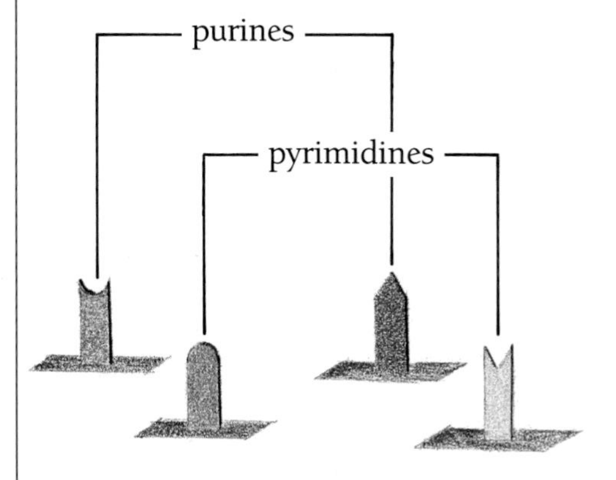

Figure 2-1. *Bases. There are two types of bases, purines (guanine and adenine) and pyrimidines (thymine and cytosine).*

consecutive bases, called a **triplet**, code for each specific amino acid in a protein molecule. In this way, the genetic messages of a species (instructions that determine how an organism looks and even behaves) are "written" out. Sections of the genetic message are carried to the various "work-stations" of the cell by a slightly different code, which might be thought of as a "dialect." The RNA code uses a uracil base (U) in the place of the thymine (T) of the DNA code. Nevertheless, it transfers the messages of the DNA with amazing accuracy. Both codes are redundant, however, which means that some amino acids are coded for by more than one triplet. The entire genetic code for RNA is given in Table 2-1.

The chemical components of DNA are identical from organism to organism, but the information contained in the base sequences of DNA varies from one organism to another. In addition, not all of the DNA codes for the organism, although noncoding DNA is not well understood, and some regions of DNA originally thought to be noncoding have turned out to perform functions. Taken together, the total DNA for a given organism is known as its **genome**. It seems hard to imagine that a single strand of human DNA is about 50 million times longer than it is wide, but it's true! A train ten feet wide, of comparable proportions, would be about a hundred thousand miles long!

Table 2-1 ***The genetic code for RNA. 64 possible triplets can be formed from the 4 letters, or bases. The code's redundancy is seen in that all 64 triplets code for 1 of 20 amino acids.***

Those with central U		Those with central C		Those with central A		Those with central G	
UUU	Phenylalanine	UCU	Serine	UAU	Tyrosine	UGU	Cysteine
UUC		UCC		UAC		UGC	
UUA	Leucine	UCA		UAA	Stop	UGA	Stop
UUG		UCG		UAG		UGG	Tryptophan
CUU		CCU	Proline	CAU	Histidine	CGU	Arginine
CUC		CCC		CAC		CGC	
CUA		CCA		CAA	Glutamine	CGA	
CUG		CCG		CAG		CGG	
AUU	Isoleucine	ACU	Threonine	AAU	Asparagine	AGU	Serine
AUC		ACC		AAC		AGC	
AUA		ACA		AAA	Lysine	AGA	Arginine
AUG	Methionine, or Start	ACG		AAG		AGG	
GUU	Valine	GCU	Alanine	GAU	Aspartic acid	GGU	Glycine
GUC		GCC		GAC		GGC	
GUA	Valine, or Start	GCA		GAA	Glutamic acid	GGA	
GUG		GCG		GAG		GGG	

Biological Information

Scientists who explore the biochemistry and structure of various organisms are impressed by their astonishing complexity. Consider, for example, that most of the bacteria studied contain a genome of some three million base pairs. For the much studied E. coli bacterium it is 4.7 million! The information in the sequence of its base pairs is so important to the understanding of the organism, that Cambridge biochemist Frederick Sanger received his second Nobel prize for determining these base sequences for a bacteria-attacking virus called a bacteriophage.

The protein-assembling information in DNA is recorded in its triplet-coded message, ready for use in transcription (see Figure 2-2 or your main text about this process). The way the information is recorded is in the successive bases along the DNA chain. Taken three at a time, these bases code for the specific sequence of amino acids in protein as it is being synthesized. The unique coding information for each species, in its entirety, is called its **genotype**. Section by

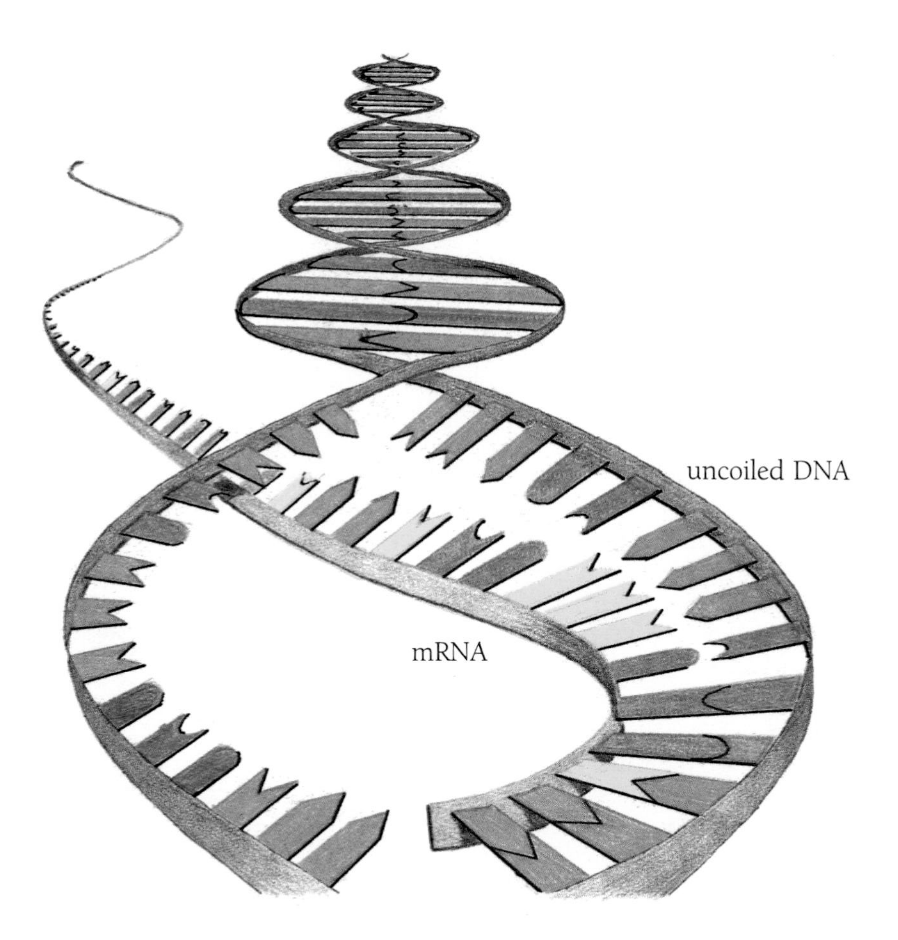

Figure 2-2. *In transcription, the sequence of bases in the DNA, which comprise its genetic message, are recorded in mRNA, so this copy of the cell's protein-building plans can be taken from the nucleus to the site of protein manufacture in the cell. During transcription, the DNA's double helix temporarily uncoils so the mRNA can form by matching each base to its complementary base as the strand of mRNA forms.*

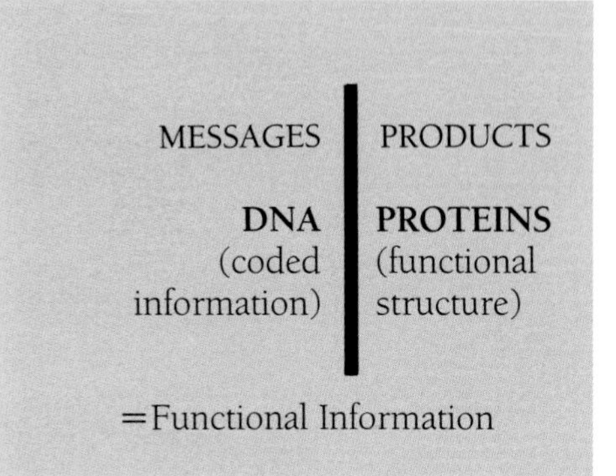

Figure 2-3. *The left side of the diagram is the messages side; the right side is the products side. The DNA carries a detailed plan, in the form of coded information, for the protein with its functional structure. Thus the DNA's protein-specifying information is functional information.*

section, its information is transferred to mRNA (messenger RNA), which carries it from the nucleus to the site of protein manufacture in the cytoplasm. Here, it ultimately directs the building of each protein out of just the right combination and sequence of 20 naturally-occuring amino acids. The resulting amino acid sequences in the newly manufactured proteins, in turn, determine both the orchestrated development and integrated structure of the organism. But where did these sequences come from? Notice Figure 2-3. There we see, on the "messages" side, DNA with its coding information, which produces functional structures on the "products" side in the resultant protein.

Is it accurate to describe the coding sequences of base symbols in DNA as "messages?" If it is, they are comparable to what is written on this page. Cell biologist E.J. Ambrose said the cell can tell us the difference between a random series and a message: "There is a message if the order of bases in DNA can be translated by the cell into some vital activity necessary for survival or reproduction."[1] Few of the 10^{839} other alphabetical sequences possible on this page would convey the same message as this one, and it is much the same with the coding sequences of DNA. We saw earlier that the principle of uniformity states that like effects (past and present) have like causes. Today, there is a growing recogition among scientists of the dramatic implication that the principle of uniformity holds for the origin of **functional information**. This is not an argument *against* Darwinian evolution. *It is, however, an important scientific inference in favor of the intelligent origin of genetic messages.*

The Hardy-Weinberg Law

One of the most important questions for both proponents of design and Darwinism concerns the behavior of the genetic world over time. Is it a world of stability? Do traits and the genes that produce them remain relatively unchanged in expression and frequency from generation to generation? Or is it a world of constant change where traits may be easily modified, lost, or altered in the frequency?

Mendel demonstrated that the units we now call genes, which determine the inheritance of traits, are stable and retained from generation to generation. For example, when Mendel crossed a pea plant having wrinked seeds with one having round seeds, all of the offspring in the first generation had round seeds. Was the gene for wrinkled seeds then lost or changed? Not at all, because in the second generation it reappeared in one-fourth of the plants. The gene for wrinkled seeds was present even in the first generation; it was merely suppressed by the **dominant gene** for the round trait.

One misconception held by many biologists during the late 19th and early 20th Century was that traits determined by dominant genes would become more and more common, while traits determined by **recessive genes** would become less common. In other words, it was thought that the frequency of a trait (and of the gene which produces it) is constantly changing in a population.

What better explanation for Darwinian change than such a shifting in the frequencies of genes and corresponding traits? This question was studied independently by Godfrey H. Hardy, an English mathematician, and Wilhelm R. Weinberg, a German physician. To pursue this question Hardy worked out the effects of random mating on the frequencies of individual traits (and thus members of gene pairs) in large populations. Here in Hardy's own words is what he concluded:

> In a word, there is not the slightest foundation for the idea that a dominant character should show a tendency to spread over a whole population, or that a recessive should tend to die out.[2]

Parental Generation

freg. of M = 0.6 freq. of N = 0.4

X

1st Generation

.36MM .48MN .16NN

freq. of M = 0.6 still freq. of N = 0.4 still

X

2nd Generation

.36MM .48MN .16NN

freq. of M = 0.6 still freq. of N = 0.4 still

Figure 2-4. *The reproduction of a parental generation having the initial gene frequencies of M=0.6 and N=0.4, with the resulting two generations of offspring. Each generation shows the same gene frequencies again, and the frequencies of the three genotypes are also stabilized at the same genotype frequencies as in the first generation of offspring. This result is what is expected in natural populations subject to the Hardy-Weinberg Law, in the absence of other factors.*

Stated differently, Hardy concluded that there is not only stability in the nature of a gene itself, but also stability in the frequency of its occurrence from generation to generation.

Weinberg's study brought him to the same conclusion. The Hardy-Weinberg Law is basically a mathematical description of genes in a population (gene pool) and has nothing to do with large scale evolution. Its basic meaning is: *the percentage frequencies remain the same from generation to generation.* Hence it confirms the implication of Mendel's discoveries—that the genetic system of inheritance is normally very stable. To see how this principle works in successive generations, see Figure 2-4.

Adherents of intelligent design assume that in the beginning, all basic types of organisms were given a set of genetic instructions that harbored variation, but were resilient and stable. Without this inherent stability, a species would soon cease to exist. Mendel's principles and the Hardy-Weinberg Law support this assumption of stability. On the other hand, Darwinism requires the operation of change factors to produce new traits as well as a certain degree of stability in order to preserve them for the origin of a species. According to Darwinists, genetic material must have some means of changing dramatically enough to result in entirely new organisms. Let's look at some of the main change factors that Darwinists cite as macroevolutionary modifiers of the functional information of a species.

Change Factors

Mutations

One observed change factor—indeed, the best known—is that of gene mutation. Mutations are not common. In fact, it has been calculated that a gene changes only once in every 100,000 to 1,000,000 replications (reproductions).[3] Another way to express the basic mutation rate is to find out how many gametes contain at least one mutant gene. Studies show one mutant gene in 10 to 100 gametes. While the cause or causes for this basic rate of mutation are not well understood, it can be increased by certain environmental factors such as heat, chemicals, and radiation.

Mutations of two fundamental classes are known to occur. The first are called *gene* or *point mutations*; these are the changing of individual genes somewhere in the DNA. (They may or may not affect the amino acids coded for by the mutated genes.) A second class of mutations are known as *chromosome mutations*, and involve not merely an individual gene of the DNA, but a section of the DNA. Chromosome mutations can result in the duplication of a segment of DNA, its loss or recombination at another place in the same or a different DNA molecule, or even its inversion within the molecule. Darwinists assume that, as we come to understand such mutations better, we will learn how they provide much of the increase in functional information necessary for macroevolution. A mutation in a coding gene, then, can be looked at as a random change in functional information. As a unit of functional information in the cell, a coding gene is much like a word (a unit of meaningful information) in a book. What do you think would happen if we randomly changed the letters in some of the words in this book? Would the book be improved? On the contrary, it is probable that random changes in the words of this book would decrease rather than increase the meaningful information they carry. If enough random changes occurred, the result would be gibberish. Mutations have the same effect in the biological world; most are harmful. In fact, only one in 1,000 is not harmful. Gene mutations can produce structural impairments and genetic diseases, to say nothing of when mutant genes are lethal. It is known that mutations produce harmful traits that natural selection actually selects against, or eliminates. Some lethal mutations, in fact, terminate the development of an individual as early as the zygote stage.

Yet if macroevolution by means of natural selection is to occur, there must first be new traits for which to select. Darwinists look to mutation as the main source of traits upon which natural selection can operate. But it has not been demonstrated that mutations are able to produce the highly coordinated parts of novel structures needed again and again by macroevolution. So far, there is no unambiguous evidence to demonstrate macroevolution's requirement of such helpful mutatins. Some examples typically given in texts are the immunity to malaria possessed by people with the sickle-cell anemia mutation, and the resistance to penicillin of mutant strains of bacteria that cause gonorrhea. In these cases it is clear that the mutations give flexibility to organisms to stabilize and maintain themselves in the presence of a severe threat, while others of their species fall victim to the invaders. What is unclear, however, is that these traits will continue to benefit the hosts when the invading organisms are gone. When an individual is heterozygous for sickle-cell anemia, the mutation gives an advantage in surviving the threat of malaria. But it does so at the expense of a mild anemia, an impairment to the function of transporting oxygen to the body's cells.

Moreover, the mutation does not introduce a new level of complexity, and it cannot be known that it is a "step in the right direction"—that it will integrate with other mutations in the future for an increase in functional information that will code for adaptations for greater complexity. Most observed mutations are harmful, and there is no experimental evidence to show that a new animal organism or even a novel structural feature has ever been produced from the raw material produced by mutations. In view of this, it is fair to remain skeptical about the claim that accidental mutations are the source of the very impressive quantities of functional information in biological structures.

Natural Selection

Darwinists reject the idea of purpose in evolution because it contradicts the allegedly mechanical, unplanned nature of the Darwinian process. Darwin justified the rejection of purpose, arguing that natural selection is a mechanism that explains what only appears to be purpose in nature. Yet he found it helpful to ascribe remarkable skill to natural selection. "Natural Selection picks out with unerring skill the best varieties" (*The Origin*). Since Darwin's time, biological literature has honored natural selection with metaphors of great artistry and skill, comparing it to a composer of music, a master of ceremonies, a poet, a sculptor, and "William Shakespeare." More recently, natural selection has become identified with the metaphor of the **Blind Watchmaker**, through Richard Dawkins' book so titled.[4] Today, however, the "creative" role of natural selection is being questioned by a growing number of scientists. Yet most of these scientists have not reconsidered the intelligent design argument which was replaced by natural selection as the supposed source of apparent design.

Natural Selection and the Fixity of Species

The idea of natural selection did not originate with Charles Darwin. Edward Blyth, one of Darwin's predecessors and proponent of intelligent design, saw natural selection as a conservative force for maintaining the **fixity** of designed species. He recognized that, if all organisms were fashioned according to basic blueprints, then there must be some kind of "quality control"—some way to keep the original types from drifting away from their original designs. He believed that natural selection played this role, as it eliminated those organisms having traits that were too deviant. Thus for Blyth, natural selection maintained the fixity of species postulated by virtually all proponents of intelligent design, including the father of taxonomy, Carolus Linnaeus. The concept, however, seemed to draw support from the notion of ideal types of Plato. This notion carried with it two aspects not held by most modern proponents of intelligent design. The first was a rigidity, a belief that no change was possible. In more recent times, intelligent design proponents have noted variation within a given level of complexity and acknowledged that change does occur but change within limits. Also rejected by most modern intelligent design adherents is the idea that the design of each and every species was perfect or ideal.

Today, the conservative role of natural selection is recognized by many. One particularly interesting example comes from the work of an evolutionist, Hermon Bumpus. Bumpus collected the corpses of English sparrows (*Passer domesticus*) killed in a severe winter storm. He then collected a sample of survivors so he could compare their characteristics with those of the dead sparrows. He found that the birds that died tended to be more extreme in their physical characteristics.

They were either heavier or lighter, or in some other significant way deviated from the norm. Individuals that varied from the norm could survive under moderate conditions, but when the going got tough, only a narrow range of variation could be tolerated. It appears that there might be an optimal body type for the species, and any bird which departs too much from this type is eventually weeded out.

Proponents of intelligent design interpret Bumpus' observations as support for their idea that every species has been given an optimal body form which maximizes its function in a particular habitat.

All biologists acknowledge **stabilizing selection**, where selection can be seen to function as a weeding out mechanism to maintain the cohesion and stability of species.

Change and the Origin of New Structures

It is probably safe to say that the genes which control the relative proportions of a bird's wing are far fewer in number than the genes actually coding for the muscular and skeletal mechanism and various support tissues (blood vessels, nerves, skin, feathers, etc.) of the wing itself. The number of genes involved in all these codes could indeed be quite large. For comparison, consider that perhaps 30 to 40 genes code for the wing structure of the fruit fly. Even in the case of insect wings, any genes coding for increase in size (a change considered important by some evolutionary scientists trying to account for the origin of flight) would be a small portion of those responsible for the overall wing structure.

Researcher E. J. Ambrose, Emeritus Professor of Cell Biology from the University of London, estimated that it "is most unlikely that fewer than five genes could ever be involved in the formation of even the simplest new structure previously unknown in the organism."[5] But Ambrose used the figure of five genes to show the incredible improbability that any significant functional information would arise by chance mutations. He began by noting the rate of favorable or neutral gene mutations. It is a reasonable estimate that no more than one new nonharmful mutation will occur per generation in a population of 1,000 (most genes have a mutation frequency smaller that one in 100,000). The probablity, then, of two nonharmful mutations occurring in the same organism is one in 1,000,000. (The probability of two independent events occurring is the product of their independent probabilities; thus 1/1,000 x 1/1,000 = 1/1,000,000.) The odds of five nonharmful mutations occurring in the same individual are one in one thousand million million! (Figure this by multiplying the five independent probabilities: $10^{-3} \cdot 10^{-3} \cdot 10^{-3} \cdot 10^{-3} \cdot 10^{-3}$.) For all practical purposes, there is no chance that these five mutations will occur within the life cycle of a single organism, as any bioligist would agree.

But suppose five nonharmful mutations occured *within the gene pool* (in different individuals) of a single species, and that these occurred over time, being preserved in the **heterozygous** state. Given the recombination potential from extensive interbreeding, couldn't the five new genes eventually come together in a descendent? The Hardy-Weinberg Law states that, in the absence of selection or other outside forces, the proportions of these five mutated genes to their non-mutated counterparts in the rest of the species' population will remain the same from generation to generation. Thus the mere production of more offspring doesn't improve the odds of their coming together through recombination. Their number will only increase if the individuals carrying these mutated genes are favored by selection or perhaps by chance in small populations (genetic drift). On the other hand, if the mutations had occurred in noncoding regions of the genome, natural selection at least would not operate to eliminate them, since a gene that doesn't code for a trait won't contribute any advantages or disadvantages for survival to the organism. This neutrality would give them a greater chance to come together through the recombinations of reproduction.

Imagine if you were to calculate the chances that the different individuals carrying these separate genes would find each other among a population of, say, one million, and mate at the proper mating time and in the necessary combinations to gather all five genes in a single individual among the resulting offspring! If we could conceive that the resulting new set of five genes could potentially code for a truly new structure, thus increasing the organism's level of complexity over its parental species, perhaps we would have a potential explanation.

But we aren't there yet. Next, suppose our set of five new genes becomes fixed in an individual chromosome, where they are not expressed in the heterozygous state because the existing genes on the matching chromosome are dominant, thus "covering" them. Suppose, too, that all five are gathered in the same region of the chromosome, which would be highly improbable, but possible by various mechanisms that occur via chromosome breakage and rejoining. If they were indeed located in one area, an additional mutation of another gene could conceivably switch that region from recessive to dominant (acting as a "switch gene" for this "supergene" region). Could this then bring about the development of a new structure in an individual organism? Such examples of the mutational development of a five- or six-gene "supergene" unit on a single chromosome are now known to control multiple *color forms* of the wings of **mimetic butterflies** such as *Papilio dardanus* in Africa.[6] But no examples are known yet for complex structures.

Ambrose pointed out that we have only discussed the probability of developing a five-gene set of nonharmful mutants. He went on to say that the difficulties of this improbable coming together of five genes:

> fade into insignificance when we recognize that there must be a close integration of functions between the individual genes of the cluster, which must also be integrated into the development of the entire organism.[7]

Ambrose never returned in his essay to remind the reader that most simple structures would require several times the five genes in the example. But by applying **information theory**, he showed that the molecular requirements of new functional information underlying new structures are so demanding that no credible theory of evolution of complex new structures, such as a bird's wing, can be built on the repeated chance occurrence of improbable events like those described. His conclusion?

> We conclude therefore that recent hypotheses about the origin of species fall to the ground, unless it is accepted that an intensive input of new information is introduced at the time of isolation of the new breeding pair.[8]

Yet the level of complexity in biological organisms is stunning—far greater than could be explained by integrated sets of, say, 50, 60, or 100 genes. Organisms exhibit systems of multiple structures, which must "fit" or integrate with one another in order to function. The coding for the majority of such structures would require sets of dozens, if not hundreds or thousands of genes for each. These structures are orchestrated to work together and to adjust the organism to its environment. A biological organism is more than the sum of its individual structures; its ability to function successfully is due to an entire "adaptational package." We cannot really evaluate the proposed origin—random gene mutations or intelligent design—of these packages, without an understanding of the concept.

Natural Selection and the Adaptational Package

An excellent example of an adaptational package is found in the giraffe. In 1827, a giraffe was presented to the Museum of Natural History in Paris. Of course, people were impressed by its long neck. Folk wisdom might explain the long neck by saying that the giraffe acquired it by repeatedly reaching up to eat the highest available leaves. Its neck had become stretched, and this stretched neck was passed on to its offspring. So gradually the giraffe's neck became longer.

Jean Baptiste de Lamarck, a French biologist, believed that alteration of the environment would alter an animal's needs, and hence its behavior; this in turn would lead to an alteration of its structure. In 1809 he wrote the following in his book *Zoological Philosophy:*

> The giraffe lives in places where the ground is almost invariably parched and without grass. Obliged to browse upon trees it is continually forced to stretch upwards. This habit maintained over long periods of time by every individual of the race has resulted in the forelimbs becoming longer . . . and the neck so elongated that a giraffe can raise his head to a height of eighteen feet without taking his forelimbs off the ground.[9]

Charles Darwin also drew attention to the giraffe's neck. He said:

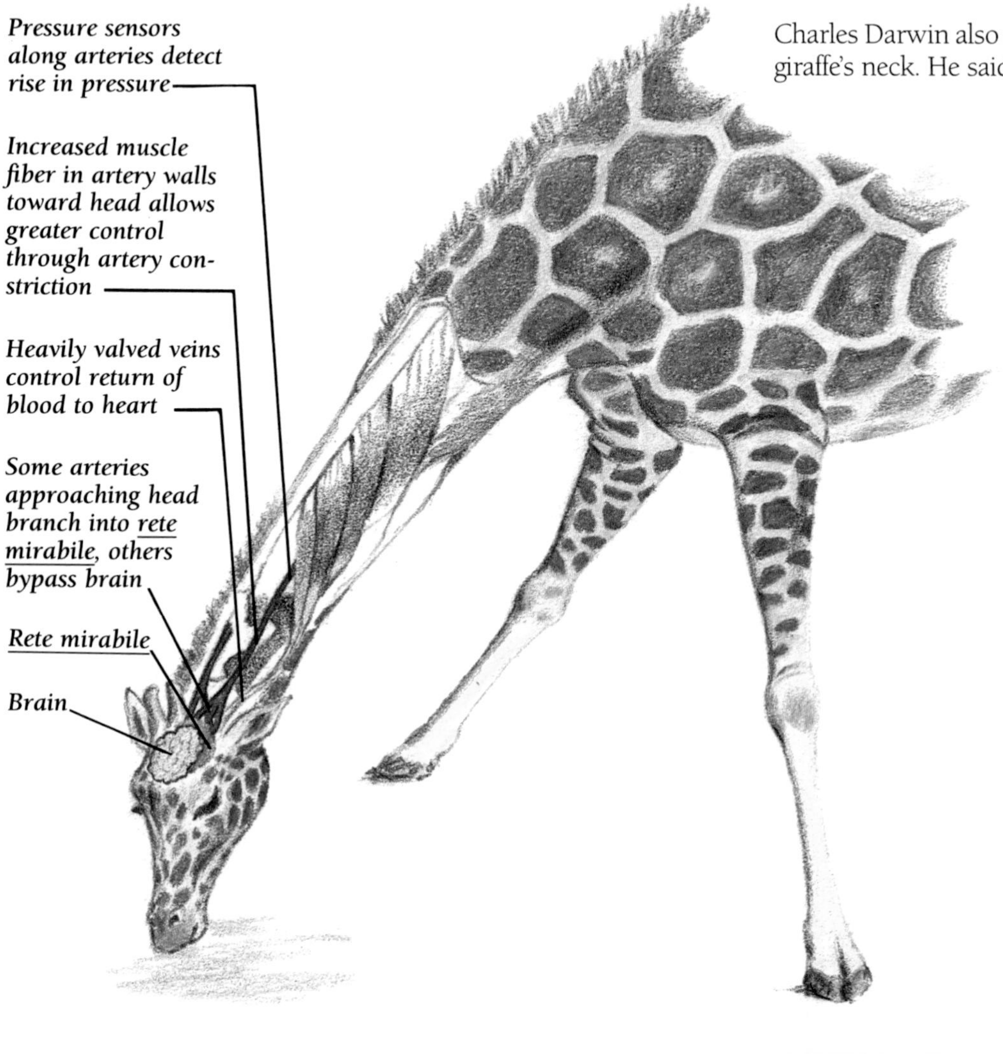

Figure 2-5. *As the giraffe drinks, an adaptational package protects it from hemorrhaging in the brain:*

Pressure sensors along arteries detect rise in pressure

Increased muscle fiber in artery walls toward head allows greater control through artery constriction

Heavily valved veins control return of blood to heart

Some arteries approaching head branch into rete mirabile, others bypass brain

Rete mirabile

Brain

> The giraffe, by its lofty stature, much elongated neck, forelegs, head and tongue, has its whole frame beautifully adapted for browsing on the higher branches of trees. It can thus obtain food beyond the reach of the other *Ungulata* or hoofed animals inhabiting the same country and this must be a great advantage to it. . . [10]

It is true that fossil giraffes have been found side by side with the fossils of sheep; the latter could have been specialized on grass, while giraffes foraged in low trees. However, it is interesting that the neck of the female giraffe is two feet shorter, on the average, than that of the male. If a longer neck were needed solely to reach above the existing forage line, then the females would have soon starved to death and the giraffe would have become extinct. Darwin was correct when he called the giraffe "beautifully adapted," but he did not have enough information to appreciate the full extent and refinement of the adaptations. Observe some giraffes eating and drinking in the zoo, and you will notice that they bend their heads to the ground to eat grass and drink water. Given their long legs, giraffes could be said to need a long neck less to reach up into the trees (which are not the only source of vegetation in many terrains) than to reach the ground to drink water.

Without the idea of an adaptational package one must think in terms of temporal priority. In order to optimally fit into its niche the giraffe needed long legs; possessing long legs it needed a long neck; and in order for it to use its long neck, further adaptations were necessary. When a giraffe is standing in its normal erect posture, the blood pressure in the neck arteries will be highest at the base of the neck and lowest in the head. The blood pressure generated by the heart must be extremely high to pump blood to the head. But when the giraffe bends its head to the ground it encounters a potentially dangerous situation. It must lower its head between its front legs, putting a great strain on the blood vessels of the neck and head. The blood pressure plus the weight of the blood in the neck could produce so much pressure in the head that the blood vessels would burst.

Mercifully, however, the giraffe is equipped with an adaptational package, including a coordinated system of blood pressure control[11] (see Figure 2-5). Pressure sensors along the neck's arteries monitor the blood pressure, and can signal activation of other mechanisms to counter any increase in pressure as the giraffe drinks or grazes. Contraction of the artery walls, a shunting of part of the arterial blood flow to bypass the brain, and a web of small blood vessels (the *rete mirabile,* or "marvelous net") between the arteries and the brain all serve to control the blood pressure in the giraffe's head. Notice that adaptations require other adaptations so that a specialized organism such as the giraffe can function optimally.

The giraffe has more than just a long neck, it has an entire adaptational package that fits it to every aspect of its way of life. Certainly natural selection works to bring out a species' distinguishing features when advantageous. This is seen best in what we are most familiar with, humans. It is easy to see in the distinguishing features of the African and the Eskimo. But proponents of intelligent design maintain that only a consummate engineer could anticipate so effectively the total engineering requirements of an organism like the giraffe.

More intriguing still is the fact that some organisms appear with adaptational packages intact at the Cambrian boundary (see Chapter 4) where multicellular life first "flowers," with no evidence whatsoever of fossil ancestors. Brachiopods from the early Cambrian series reported in scientific literature illustrate this phenomenon.[12]

An additional fascinating characteristic of some adaptational packages is that of multifunctional adaptations, where a single structure or trait achieves two or more functions at once.

The existence of such a sophisticated adaptational package is taken as evidence by the proponents of intelligent design of their theory. In our experience, only an intelligent designer has the ability to coordinate the design requirements of multifunctional adaptational packages. If this is true, asks the design proponent, how can the blind, chance forces of nature excel in achieving the technical feats that confer distinguished status on successful engineers? Multiple accidental gene mutations are a highly improbable source of new genetic information to code for multi-functional structures.

Populations of moths that over generations shift in color from light to dark, or mosquitoes that exhibit resistance to DDT are often cited as examples of evolution by natural selection. Such relatively small changes in the gene frequency of populations, whether involving mutations or not, are called microevolution. Yet design proponents ask if such minor changes really constitute evidence that major changes involving adaptational packages took place; many are skeptical about this conclusion. A shift in the dominant moth coloring would require no new genetic information, because the **alleles** are already present in the population. In contrast, major changes would require major coordinated adaptations, which would in turn require impressive amounts of new functional information. When we fully appreciate the informational requirements for the origin of even a minimal new structure, much less the origin of a major adaptational package, we can see what a "tall order" such origins would be for the "blind watchmaker."

Natural Selection and Gene Combinations

Darwinism attempts to account for adaptational packages by gene mutation and natural selection. Ambrose argues that the **selection pressure** from the environment is too general for the demands of evolution:

> But the sort of message which the physical or biological environment can transmit to the organism in the way of new information is an extremely simple one, of the yes or no type such as "Can I find food higher up the hill or not?"[13]

Such simple information, however, is not sufficiently constraining to "pick out with unerring skill" from the mutations available and build complex new structures. Natural selection is at work helping species to adapt better to changing environmental conditions, but not in the way that many think. Natural selection is alleged by Darwinists to select *new* genes, thereby producing *new* organisms, a view for which there is no empirical evidence. In reality, as sexual reproduction "reshuffles the genetic deck," natural selection operates on various combinations of already-existing genes in response to environmental demands. Let us look at some examples of natural selection at work.

The English sparrow was introduced to North America in 1850. Some people wanted to establish this species in America, but their early attempts failed. It is interesting that they wanted to introduce this bird, since it later became a nuisance and could not be kept from multiplying. Eventually, the birds obtained a limited foothold in a few localities where they remained relatively few in numbers for several years. Ultimately, they adapted to their new habitat and their population underwent runaway growth (probably due to the widespread use of the horse, which provided a food source of insects that bred in the horse droppings and of grain used for horse feed). Today English sparrows live throughout most of the continental United States.

Advantageous Gene Combinations

This success story has an interesting footnote. When samples of the sparrow population from several geographic areas within the United States were taken, it was discovered that the birds from colder climates were, on the average, larger than those from warmer climates, and also had shorter extremities. Apparently the ideal sparrow body type varies according to geographic region (see Figure 2-6). These differences of size and extremity were long known among similar species of animals living at different latitudes. But with the sparrow it had now been observed within a single species, raising a different question: How can this geographic variation within a species be explained?

The answer lies in the way some genes express themselves. Mendel worked with genes that were dominant and recessive. Both alleles were present all along, but only one or the other for a given trait was expressed. There are other genes, such as those for human skin color, where many gene pairs affect the trait expression in an additive or cumulative way. In other words, the color of skin is determined by the number of dark and light genes work-

Figure 2-6

Figure 2-6. *The pattern of distribution of the English sparrow by body types illustrates that species adapt to environment as advantageous combinations of genes already present are utilized.*

Figure 2-7. ***Skin color in humans is determined, not by just one, but a combination of genes. Thus the skin colors of a couple's children may be very different from their own, depending on which of the skin color genes are expressed in each.***

ing in combination. Two completely hybrid individuals could produce offspring exhibiting the complete range of possible skin colors (see Figure 2-7). In such a distribution, the extremes of light and dark would be the rarest. Most of the population would tend to center around the average between the extremes. Notice that this differs from the mistaken blending idea in that here we are dealing with a *number* of dominant genes that are preserved from generation to generation. It is the combination of genes that determines skin color.

If this population showing the entire spectrum of skin shades moved into a geographical area where dark skin was an advantage, what might happen? Natural selection could select the darker skinned individuals and thus the combination of genes producing the darkest color. Would there be a change from one species to another? No. The only change would be in the frequency of certain gene combinations.

In the case of the sparrows, the first birds to come to North America, known as the founders, likely possessed all the genes necessary to produce the most extreme examples of body and extremity size observed today. But they probably did not have the genes combined in the same way as do the more extreme forms of today; they had not developed specialized combinations. This is why the sparrows became established at first only in small populations, and time for specialized combinations of genes was needed before they became vigorous and abundant. When such combinations occurred, they tended to place those individuals at an advantage, and were favorably selected by regional environmental factors. Yet it was the combination that was advantageous, not the genes themselves.

Obviously then, genetic diversity in itself may be an advantage to a population. It has been suggested that the diverse range of antibodies in the human gene pool, which are differing proteins (immunoglobulins), may make it difficult for any variety of bacterium to become established in the human population as a whole.

As illustrated by the moths and sparrows, much of the variety we see within species does not indicate even microevolution at work, but of the expression of new combinations of genes that were already present. The term describing populations with the genetic diversity for several different anatomical and morlogical variations is **polymorphic**. Combinations of both expressed and unexpressed genes (in the genetic reserve) can give biological populations adaptive potential. Even among a relatively small population of founders there is much greater polymorphism (potential diversity in body form) than appears to exist at the beginning. Let us look at another example.

Organisms living in cold climates tend to be bulkier and to have shorter extremities than those living in warm climates. This difference reflects the relationship of surface area to mass. A body will radiate heat more rapidly if its ratio of surface area to volume is high (increased, for example, by a slender body form with less mass and long and skinny appendages). It will conserve heat if its ratio of surface area to volume is low (decreased by a round body form with short stocky appendages).

An Eskimo and an African from the Nile region belong, of course, to the same species, but the contrast in their body builds is striking. The long arms and legs of the African, an advantage for radiating excess body heat in the sunny tropics, would be a disadvantage in the arctic, where they would cause excessive cooling and be more susceptible to frostbite than would the shorter extremities of the Eskimo. Natural selection selected different body types to adapt the same species to a wide range of climatic conditions. This should not be confused with the transformation of one species into another; there has merely been an adaptation of a single species to a variety of environments. It is gene combinations, not individual genes, that are subject to natural selection, for it is they that determine the totality of the organism's body type.

Natural Selection and Genetic Diversity

If a population is unable to expand into new environments or adapt to changing environmental conditions, what would happen to it? First of all, it would probably remain small, and small population size is a danger to any species. When an organism mates it contributes a sperm or an egg to the offspring. The gametes contain only half of the organism's genes. When mating occurs the partners contribute only half of their total complement of genes (excluding sex-linked alleles). This is compensated for by having a large number of offspring, because the same half will probably not be contributed each time. The larger the offspring population, the greater the number of gene combinations and the greater the percentage of gene pool preserved. It follows that a low reproductive rate will increase the probability that genetic information will be lost. If this loss of information occurs, it reduces the polymorphism of the population. If it continues, the species' ability to adapt to changing environments will be lost and the species itself will become extinct. In the case of rare or endangered species today, populations could easily become extinct due to a loss of genetic diversity (or variability) occurring in this way.

The healthiest thing for a species as a whole is to reproduce randomly. This insures that its hidden adaptive genetic potential is preserved. The genius of natural selection is that it serves as a force, not to change species, but to preserve them. This it does by selecting gene combinations that allow a species to survive in new or changing conditions. It thus insures that a species will be able to flourish and increase in numbers.

If natural selection is a conservative and not a creative force, then where did species come from in the first place? The various answers given to this question will be examined in the next chapter.

Suggested Reading/Resources

The Mathematical Foundations of Molecular Biology, by Hubert P. Yockey. New York: Cambridge University Press, 1992. An application of information theory to the subject of information in molecular biology.

References

1. E. J. Ambrose, 1982. *The Nature and Origin of The Biological World.* New York: Wiley, Halsted, p. 26.
2. G. H. Hardy, 1908. *Science* 28, 49-50.
3. T. Dobzhansky, 1951. *Genetics and the Origin of Species.* New York: Columbia University Press, p. 59.
4. R. Dawkins, 1987. *The Blind Watchmaker.* New York and London: W. W. Norton.
5. Ambrose, p. 120.
6. E. B. Ford, 1971. *Ecological Genetics.* London: Chapman and Hill.
7. Ambrose, p. 123.
8. Ibid., p. 143.
9. Jean Baptiste de Lamarck, 1914. *Zoological Philosophy.* Translated by Hugh Elliot. London: Macmillan, p. 122.
10. Charles Darwin, 1859. *The Origin of Species.* 1963 Reprint, New York: Heritage Press, p. 181.
11. R. H. Goetz, et al., 1960. *Circulation Res.* 8, 1049-1058; and W. E. Lawrence and R. E. Rewell, 1948. *Proc. Zool. Soc.* London 118, 202-212 (see esp. p. 210).
12. R. A. Raff and E. C. Raff, eds. 1987. *Development as an Evolutionary Process.* New York: Alan R. Liss, pp. 71-107.
13. Ambrose, pp. 140-141.

EXCURSION CHAPTER 3

The Origin of Species

Introduction

How the species originated is an ancient question, one that people have tried to explain for thousands of years. When Charles Darwin first published his book, *The Origin of Species*, it received many criticisms. One of the criticisms was that he had not given his predecessors sufficient credit for their work on the problem. Darwin responded to this criticism in later editions, where he listed a large number of supposed evolutionary thinkers who had preceded him. If one takes time to check, it is obvious that many of these men, such as Ovid and Empedocles, were not thinking of evolution as we understand the term, but a crude form of spontaneous generation.

A common explanation of origins in most ancient cultures was creation by the gods. On closer inspection we see most of these ancient creation myths were personifications of nature. Even so, in the same data from antiquity, modern views of intelligent design and macroevolution find their ancient roots. Design proponents point to the role of intelligence in shaping clay into living form. Evolutionists, on the other hand, point to the clay itself as the stuff of which life is spontaneously generated by nature—stuff which most of the time was personified as a god. Those two alternative concepts of origins thus have long histories extending from ancient times to the present.

It is important that we have in mind today's views of intelligent design and macroevolution, otherwise, we wind up dismissing one or the other for holding wornout and discarded notions of the past. For example, it is unfair criticism to say adherents of intelligent design don't believe that species change, or to say that Darwinists believe a monkey evolved into man. Whatever past followers of intelligent design and Darwinism might have held, it is clear that modern proponents do not hold such views.

There is, however, a central core idea that modern adherents of each view hold in common with their forebears. Materialists through all the ages have held the concept of life emerging from simple substance. What the substance is, and the mechanism of emergence, are details that have changed to characterize many different theories of evolution. Likewise, proponents of intelligent design throughout history have shared the concept that life is like a well-crafted object, the result of intelligent shaping of matter.

With respect to the existence of a single, unified modern theory of intelligent design, holders of this view point out that, while there are difficulties that need to be worked out, all adhere to the same fundamental aspects. Most significantly, all design proponents hold that major groups of organisms had their own origins. While there is diversity among design proponents, it is not unlike the diversity among Darwinists with respect to modern evolutionary theory.

What unit of classification was originally designed? Was it the species, or genus, or family, or even the phylum? Some Darwinists insist that those who hold true design must be able to answer this question. In point of fact, the question is irrelevant, because categories of classification are largely artificial, human groupings. This, in part, contributes to the dispute over how to define species. Both design proponents and Darwinists have had difficulty giving a consistent definition of species. But despite this difficulty, a species is widely considered to be an inter-breeding population, the offspring of which are fertile. Indeed, it is surprising that taxonomists agree as much as they do. There are minor invertebrate phyla which contain not more than a dozen species. It is theoretically possible that a phylum and a species can be identical; a phylum could contain only a single species. An example of this is the *Caryoblastea* (KAR-ee-o-BLAST-ee-ah) in the phylum *Pelomyxa palustris.* In any case, there can be no doubt that if a designer produced phyla initially, those phyla must contain specially fashioned representatives, or species.

But would the original species have remained as they were when first fashioned? Influenced by Plato's concept of unchanging essences, or types, some 18th and 19th Century proponents of intelligent design believed that not only were species unchangeable, they were also inextinguishable. Today, holders of design accept the idea that species can change within limits, and recognize that many species have become (and are becoming) extinct, too. So change is accepted by modern adherents of design; the challenge is discovering the extent of change that takes place, the nature of the relationship between living and extinct species, and, most important of all, the limits of change.

This chapter extends the discussion of genetic change begun in Chapter 2. However, the focus of this chapter will shift from genetic patterns to the environmental factors that might influence them. Over the years, numerous investigations have explored the questions of whether species are "infinitely plastic," capable of unlimited change, or whether change is limited. Darwin advocated the unlimited change view. The accumulated evidence to date, however, severely questions Darwin on this. For example, the Bumpus study of birds (Chapter 2) showed a remarkable tendency for birds to vary within limits. Hermann J. Muller labored for many years conducting mutation experiments with the fruit fly *Drosophila* to demonstrate unlimited change, and found the same tendency: change occurs only within definite limits. Others have tried, as well. Such attempts have all met with uniform lack of success, and ultimately died a quiet death. Hardly anyone is still trying to furnish an observable basis for Darwin's view of unlimited change. The peppered moth, *Biston betularia,* in England shows a genetic capacity for changing color, adapting the species to a changing environment. But no evidence suggests the black variety is anything but a variety of the original species. Selective breeding experiments have resulted in better tasting beef, "super chickens," and increased protein content in corn. In general, domestic plant and animal production has increased substantially through exploiting genetic variation. As experience has shown, however, the variation eventually ceases, and further change is not possible.

The Darwinist, however, believes species have unlimited potential for change even if scientists have not been able to experimentally produce it. Darwinian theory holds that the diversity of contemporary species arose through descent from a common ancestor. According to Darwinists, we must regard lack of experimentally induced, unlimited change as a problem in need of research, not a basis to doubt macroevolution. This illustrates well the fact that proponents of intelligent design and Darwinists are divided over perspective or viewpoint of interpretation, not data. Design proponents agree that research should continue, but maintain that empirical observations like those above presently provide clear support for their view of limited change.

Allopatric Speciation

The Concept Explained

Although experimental verification of unlimited change of species seems out of our grasp, Darwinists are hopeful that a way will one day be found based on their interpretation of the fossil record. Meanwhile, they point to presently observed speciation as a visible occurrence of microevolution. The famous geneticist Dobzhansky said, "it is no exaggeration to say that if no instances of uncompleted speciation were discovered the whole theory of evolution would be in doubt..."[1]

Several mechanisms for generating species have been advanced. The most common view of speciation held by Darwinists today is termed **allopatric.** This refers to the theoretical situation in which a single population becomes geographically divided into two populations (see Figure 3-1). These populations cannot interbreed, so it is theoretically possible for different mutations to accumulate in the two populations. The result is that they supposedly diverge both genetically and physically until two species exist instead of the original one. Major evolutionary change could occur only when the two populations that founded the diverging lines become genetically isolated from one another. In other words, although speciation may not be followed by major evolutionary change, speciation would have to be the first step leading to such large-scale change.

How do Darwinists believe speciation might occur? Suppose that a single species of fruit fly inhabited a forest and was geographically restricted to it. Suppose further that lumbering separated the forest into two distant regions so that the flies were also separated into two subpopulations unable to interbreed because of geographic isolation. According to Darwinists, this situation could lead to two new, separate species of fruit flies.

An event such as lumbering could produce isolated populations of considerable size. However, small subpopulations could also result due to a variety of genetic isolating phenomena. Let us examine the theoretical mechanisms of genetic change in small populations.

Genetic Change in Small Populations

Genetic change is likely to take place more rapidly in small populations than

Figure 3-1. Geographical barrier. A small subpopulation may be cut off from its parental population resulting in the elimination of much of the original gene pool and major changes in gene frequencies.

in large ones. One reason for such rapid genetic change is that members of small groups are forced to interbreed very closely. Even a neutral mutation (one that is neither harmful nor beneficial) could become very widespread in a population of, say, a small, isolated village.

Another factor that influences genetic change in small groups is the random way in which genetic traits are distributed. The frequency of genetic traits in a small group often "drifts." This is understandable from what we know of the laws of probability. Suppose there are two gray guinea pigs of opposite sex, each of which has one albino parent. From our knowledge of genetics we can assume that these animals are heterozygous, that is, Gg (assuming that G stands for the dominant gray trait, and g for the recessive albino trait). Now suppose these gray guinea pigs mate and produce a litter of four offspring. How many of the offspring will be gray and how many will be white?

Each parent guinea pig (having a Gg genotype) will produce two kinds of gametes, one containing G and another g. On the average, the random combination of these gametes (see Figure 3-2) should produce three gray offspring (GG, Gg, and gG) and one albino (gg). Theoretically, this is what should happen, but in the real world it might not. All of the offspring may be gray, or one half gray and one half white, or some other combination. Why do such deviations from expected ratios occur?

According to the laws of probability, the smaller the number of events, the larger will be the deviation from expected results. For example, when you flip a coin you have a 50:50 chance of turning up heads. If you flip a coin four times you expect to see a total of two heads and two tails. Does this always happen? Try it and see. You may even find that you flipped four heads in a row, or four tails. The reason you can get such a large deviation from expected results is the extremely small number of trials. If you could toss your coin ten thousand times you would find that results would be very close to 50:50. The larger the number of events or trials, the closer the results will come to the expected probability. On the other hand, the smaller the number of events, the more likely that the results will deviate from the calculated probability.

This same principle of deviation can be applied to our guinea pigs. Just about any percentage of colors could occur in the offspring because of their small population size. Such chance changes in gene frequences are called **genetic drift.**

When changes in the frequency of genes are extensive enough, one allele of a gene pair can disappear completely from a population. This does not normally occur in large populations, but, like genetic drift, is very common in small ones. The process of establishing a gene by the loss of its allele is called *fixation.*

Fixation is likely to occur in a small population because of the large effects of random events. Mathematicians use the term "drunkard's walk" to picture the effects of random events. If a drunken man staggers down a street in a series of random directions, he cannot continue walking indefinitely. Sooner or later he will crash into something such as a wall or a lamppost. Similarly, in a small population random genetic events will eventually eliminate one of a pair of alleles unless some other force, such as natural selection, preserves that particular gene.

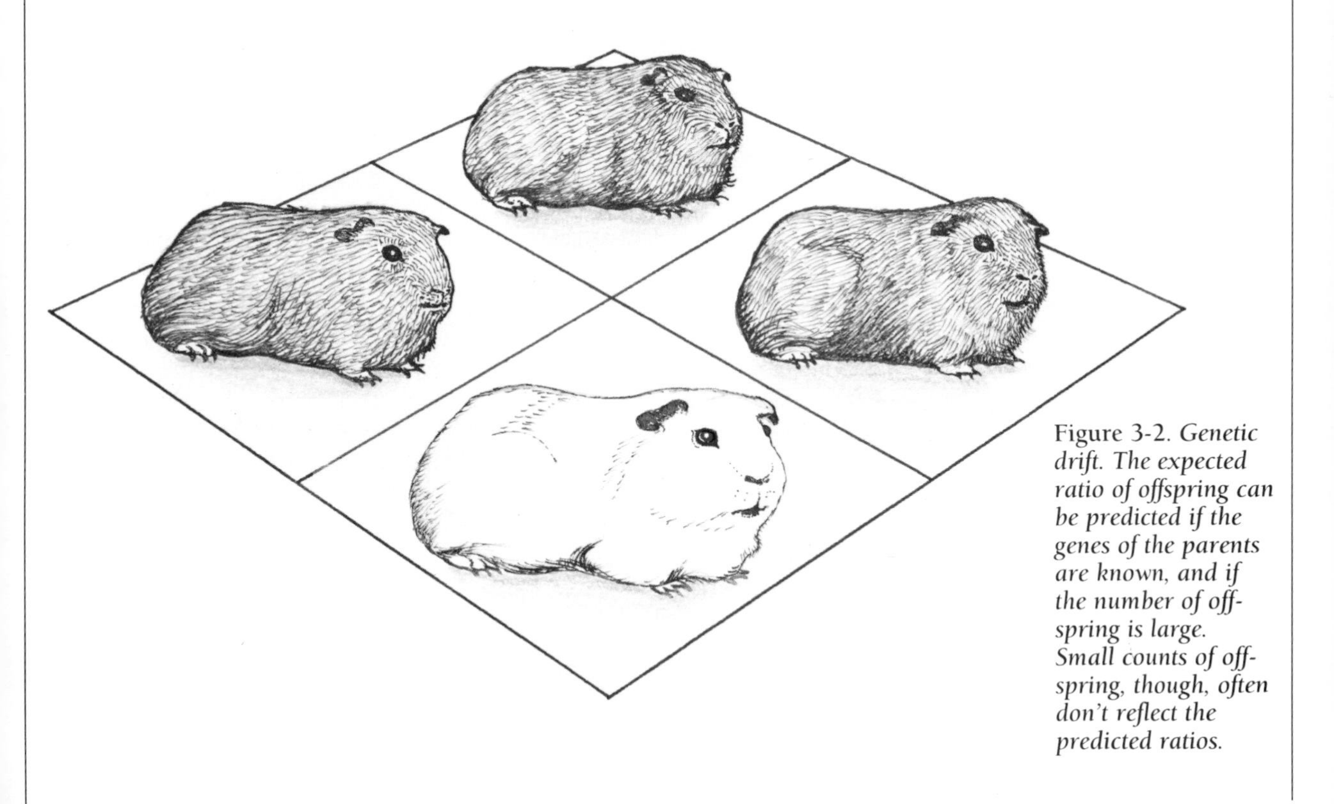

Figure 3-2. *Genetic drift. The expected ratio of offspring can be predicted if the genes of the parents are known, and if the number of offspring is large. Small counts of offspring, though, often don't reflect the predicted ratios.*

The bottleneck effect (A) occurs when some natural event kills most of a population, thereby reducing the variation within the species' gene pool. The founder effect (B) occurs when two or a few founders migrate to a new habitat, thus reducing the variation within the new gene pool of the daughter population.

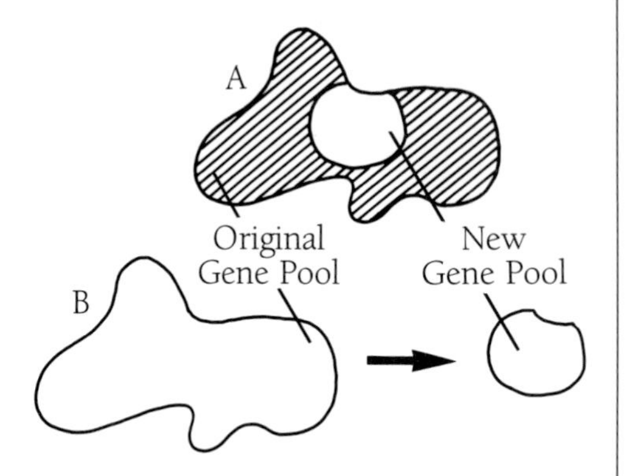

In such a situation, natural selection is not a force for change, but for preservation. Notice that the effect of genetic drift and fixation is to produce a loss of genetic diversity. This occurs when one of a pair of alleles coding for some characteristic is lost, resulting in the loss of genetic information.

The **founder effect** can also profoundly affect gene frequencies. This effect operates when new habitats, or ecological niches, are colonized by a very few founder organisms. A small sample of a population, as represented by a founder group, is not likely to have the same gene frequencies as the larger group from which it comes.

As an example of the founder effect, suppose that our four guinea pig offspring were brought by settlers to a remote island where two of them escaped to become the founding population. If both were GG, as could easily happen, the frequency of G alleles in the founding population would be very different from their frequency in the original population. In such an event, notice how great the change would be: the G gene has increased from a 50% frequency to 100% in the newly isolated population. This is a much faster rate of change than could ever have been produced by mutation or natural selection.

Similar to the founder effect is the **bottleneck effect**. This occurs when some environmental event such as glaciation (in the possible case of Neanderthal people), drought, or even a severe winter, greatly reduces the size of a population. The gene frequencies in the remaining small population can vary greatly from those of the original population.

Reproductive Isolation

Can the four factors we have studied (genetic drift, fixation, founder effect, and bottleneck effect) contribute to species formation? The answer is yes. Consider the Hawaiian fruit flies. These flies have diversified into a number of sibling and near-sibling species mainly by genetic drift and founder effect. Here is what probably happened. In fruit flies, mating depends on the exact performance of an inborn courtship behavior pattern. All elements of the ritual must be performed with precision in order for mating to follow. As you might guess, this complex series of behaviors is under the control of various genes. If a male fruit fly is missing one of the genes controlling the ritual, he will fail to perform the part controlled by that gene. If this male tries to mate with a female from the ancestral population, she will have nothing to do with him. If the male is able to find a female to mate with who also lacks the same gene and will accept him, a new species can become established.

This same type of phenomenon has been observed *within* species such as salamanders, frogs, sea gulls, and fruit fries. A certain species of butterfly, *Heliconius erato,* (hel-uh-KON-i-us eh-RAH-to) living in the Amazonian and Central American rain forests provides a classic example. This is an immense jungle region which, in its original state, once stretched over an area the size of the European continent. At one time, the species of butterfly we are considering had a continuous distribution over this vast region. However, numerous subpopulations now exist within this group. These subpopulations have striking but highly varied wing markings. Although adjacent subpopulations can interbreed with each other, subpopulations found at the opposite edges of the rain forest have less interfertility, and could, over time, lose their ability to interbreed.

The situation is similar to that of the dog species which contains many subpopulations (in this case, breeds) such as Great Dane, Boston Terrior, and Chihuahua. The range that prevents interbreeding among the dogs is a range of size. The breeds at the extreme ends of the size range cannot interbreed with each other, yet they are considered part of one species because they can interbreed with intermediate size breeds.

In the case of the butterflies, those at the extreme edges of the geographical range can interbreed with adjacent populations, and are ultimately connected to each other by a "breeding chain" (see Figure 3-3). In theory, a mutant gene originating in the extreme north of the range could, over time, spread through the breeding chain to a population in the extreme south. All *Heliconius erato* subpopulations are considered to be members of the same species. If some environmental event, say the development of a great plain, cuts the jungle in two, then the intermediate links

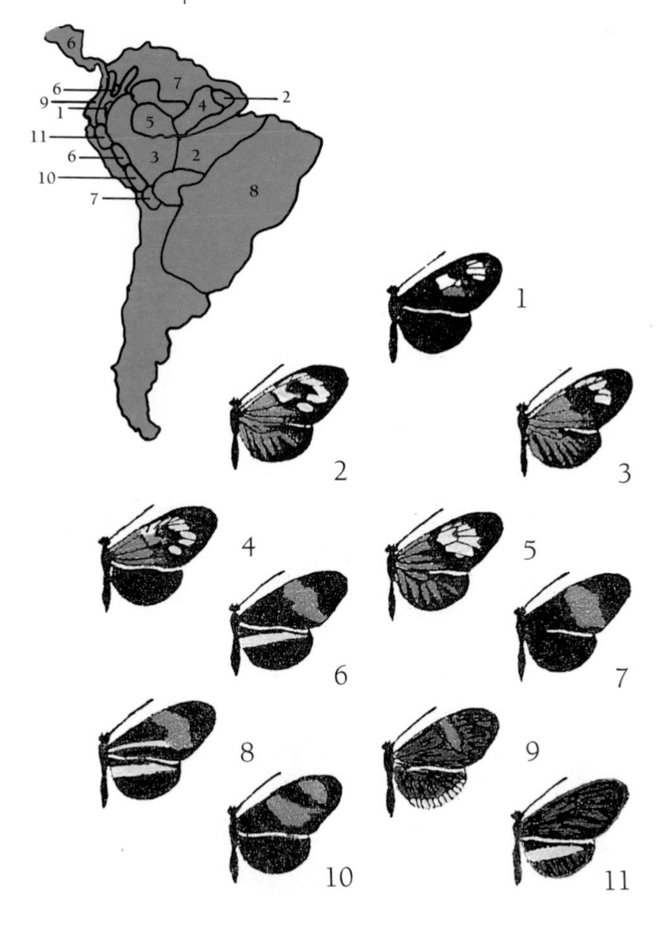

Figure 3-3. *The distribution of Heliconius erato. The numbered zones indicate the distribution of the subpopulations of Heliconius erato shown. All adjacent subpopulations can interbreed with one another.*

would no longer exist; because of reproductive isolation, the extreme subpopulations would now be considered "sibling species."

Or consider the case of the two kinds of squirrels separated by the Grand Canyon. The Kaibab squirrel inhabits the north side of the Canyon, while the Abert squirrel inhabits the south side. It seems evident the two descended from one original population. Rarely, however, can squirrels from both populations come together, and thus there is no interbreeding between them. For some time biologists have disagreed as to whether the squirrels had reached the level of two separate species, or if they remain one species with two well-marked varieties (subpopulations). It is easy to speculate that the small groups of founders exhibited different gene frequencies from the ancestral population. Then, the different environmental conditions on the two sides of the Canyon could also have selected certain genotypes at the expense of others. The combined result is a "new" variety, or perhaps species, of squirrel.

Can reproductive isolation (with resultant speciation) occur without geographic isolation? Consider the following. Not all "jungle" is the same. Areas within a given rain forest can have significant changes in environmental conditions due to varying altitude, soil types, soil water, species of plants, etc. No doubt the nature of the "best fit" between genetic makeup and the support of the environment could vary over time (see Figure 3-4). In other words, natural selection could eliminate individuals carrying certain genes so that the subpopulation in an area becomes confined to an environmental niche and, hence, reproductively isolated from adjacent subpopulations.

Figure 3-4. ***Subpopulations of many insects inhabit specific trees, which can bring about reproductive isolation from nearby subpopulations.***

Do Losses Lead to Gains?

What are the implications of these gene frequency changes in small populations? According to Darwinism, once reproductive isolation occurs, the road to large-scale evolutionary change lies open, and out of isolated populations new genera, families, orders, and even phyla can emerge. The first step, according to this theory, must be reproductive isolation, and then the way is cleared for organisms as diverse as earthworms and human beings to eventually develop from the isolated populations.

Design proponents view reproductive isolation in a different light. They point out that speciation actually represents a relatively minor change in a population, and does not provide evidence that the great amount of functional information demanded by macroevolution can be acquired through genetic modification. Genetic drift, fixation, the founder effect, and the bottleneck effect discussed earlier are each based upon losses of certain genes.

The striking thing about the diversity of biological organisms, whether living or fossil forms, is the very formidable quantities of new functional information that characterize each form as one moves up the scale of complexity. As we said, Dobzhansky acknowledges the necessity of an observational basis for speciation, a process we can explain strictly by a loss of information. Of course, change that occurs through the loss of information must soon come to an end. Yet surely, the problems of speciation pale by comparison to the problems in accounting for the sources of functional information that code for the spectacular pageant of life on earth.

The Darwinian explanation is that earlier stores of information are elaborated by extensive random mutations, and the resulting "enriched" gene pools are then weeded out by natural selection to achieve new functional information. Remember, the coding sequences of nucleotide bases in the DNA cannot be just any sequences; they must be specific and meaningful ones.

The experimental work on the origin of life and the molecular biology of living cells is consistent with the hypothesis of intelligent design. What makes this interpretation so compelling is the amazing correlation between the structure of informational molecules (DNA, protein) and our universal experience that such sequences are the result of intelligent causes. This strong analogy leads to the conclusion that life itself owes its origin to a master intellect. One can talk about adding innumerable random mutations, but proponents of intelligent design still wonder: How were such impressive gains in functional information consolidated? It is a fair and crucial question.

Does speciation fit with the theory that species were originally designed? If the intelligent design explanation is true, there may be species on the face of the earth that have undergone no substantial change since their beginning. On the other hand, the idea of intelligent design does not preclude the possibility that variation within species occurs, or that new species are formed from existing populations (as illustrated by the previous discussion of squirrels). The theory of intelligent design does suggest that there are limits to the amount of variation that natural selection and random change mechanisms can produce. Design proponents are interested in research that will answer questions such as the limits of change that exist, how we can identify original species (if any) alive today, and what the exact biological definition of a species should be.

The Tempo of Evolution

Darwin understood that the existence of apparent gaps in the fossil record would prove to be the basis for one of the strongest arguments against his theory. In his *Origin of Species* he discussed the problem and credited the "extreme imperfection of the fossil record"[2] as the primary explanation.

Now more than a century has passed, but the absence of unambiguous intermediate forms linking living with fossil groups, and fossil groups with one another has persisted. This has caused a growing number to question Darwin. In order to avoid this problem, several explanations of the gaps have been advanced.

Punctuated Equilibrium

Punctuated equilibrium is the most widely discussed Darwinian hypothesis advanced to explain the existence of gaps in the fossil record. For that reason, we will take up the subject more fully in Chapter 4. Our interest in it here is as it relates to the genetic and environmental factors we are discussing. According to punctuated equilibrium, major evolutionary changes in small populations take place rapidly (say, in a few hundreds to several thousands of years) rather than slowly (that is, in millions of years) as conventional evolutionary theory usually holds. Punctated equilibrium theorists reason that when an organism becomes adapted to an ecological niche (which roughly means "life style"), the major adaptations must occur early and rapidly. If they do not occur rapidly, the organism is in danger of displacement by other, perhaps potentially better adapted, competitors. The survivors will be those which most quickly and thoroughly adapt to the demands of this new life style. After an initial burst of evolutionary change, further refinements are likely to be minor and to occur slowly. As genes become fixed, the population will exhibit less and less variability and will lose its ability to further adapt. Such a population, though too well adapted to be threatened by any potential competitor at the time, is nevertheless a "sitting duck" for extinction, should its ecological niche itself be seriously threatened.

The well-adapted dinosaurs may illustrate such vulnerability. Although no one knows why the dinosaurs became extinct, many have suggested that it came about through environmental changes to which the dinosaurs were unable to adapt, rather than as a result of competition from birds and mammals.

The first two assumptions of the punctuated equilibrium hypothesis are rapid initial adaptation to a niche and extinction due to factors other than competition. If true, these two factors might substantially account for the gaps in the fossil record. The earliest populations in a lineage of organisms becoming adapted to a new life style would, according to this theory, evolve most quickly. If the changes took place rapidly, the probability that any of the intermediate forms would have been fossilized would be low. Instead, the fossil record would most likely contain examples from the long history of the well-adapted later forms, which show little or no changes.

In addition to these two assumptions, there is a third—that rapid evolution is most likely to take place in small, isolated populations. The most rapid changes in gene frequencies are caused by random factors such as the bottleneck effect and the founder effect. In order for random factors to operate effectively in a population, the population must be small. When, for some reason, a population's geographical isolation comes to an end, it would then be genetically isolated from the ancestral population. Since the changes that result

Some think the extinction of the dinosaurs occurred because they didn't have the genetic diversity to adapt to environmental changes.

in intermediates would occur in small populations, few such intermediates would ever have existed, making it even less likely they could have been recorded in the fossil record.

Thus, the proponents of punctuated equilibrium, with its three basic assumptions, see their hypothesis as actually predicting the existence of gaps in the fossil record. They view these gaps as evidence for their view. However, there are some serious objections to the theory. Design proponents have long asserted that gaps in the fossil record are evidence for intelligent design. This assertion has brought a familiar objection; the lack of something, which gaps certainly are, cannot be evidence. Such would be an argument from silence. But surely if gaps cannot support an intelligent design, they cannot support punctuated equilibrium either.

The problem is compounded when we consider that the events which produced the fossil record are historical (singular) rather than repeatable, like the motion of the planets, for example. Such scenarios about events that are done and gone win our allegiance by being reasonable in light of the total evidence, not because they are proven. Without observation or testing, as could be done for a theory of planetary motion, there is no "proof" or "disproof" possible.

Other objections to punctuated equilibrium have originated from Darwinists themselves. For example, many have critized it for its lack of a plausible genetic mechanism. While suggestions of genetic mechanisms or factors that may help make them up have been offered, none has given an adequate account for the functional information present in all organisms. Yet many biologists say that if large evolutionary changes do occur, they would have to occur by some means at least similar to punctuated equilibrium. To its credit, punctuated equilibrium illustrates that Darwinian theory can be tentative, subject to evidence, and may change, however slowly, in light of new evidence. The question still remains, did punctuated equilibrium occur?

From their perspective, design proponents recognize that a tremendous amount of research still needs to be done on the fossil record. There is still much detailed information buried in the record that needs to be analyzed. In the years to come, design proponents hope satisfactory answers will emerge to questions such as why organisms become extinct, and what limits of biological change are reflected in the fossil record.

The Failure of Natural Selection

Ernst Mayr of Harvard once remarked, "the book called *The Origin of Species* is not really on that subject."[3] His colleague George Gaylord Simpson also stated: "Darwin failed to solve the problem indicated by the title of his work."[4] Darwinists still have not solved the fundamental problem of how species originate.

The origin of the species, or new life forms, is evident in the design proponent interpretation: they were intelligently designed by some informative selection of the material for their genotypes. From a Darwinian point of view, however, we are no closer now to understanding the origin of species than before. As we have seen, there are several macroevolutionary theories, but none today is really strong. Darwin's use of artificial selection as an analogy for natural selection made his theory persuasive. But times have changed; today there is a strong case based on experiment that there are limits to genetic variation, which diminishes the persuasive power of Darwin's argument. Moreover, a growing number of scientists accept natural selection as a reasonable explanation for the modification of traits but not for the origins of new structures.[5]

If natural selection operates on genetic variability to produce new species, then the Darwinist is faced with several difficult problems. First of all, if organisms can be modified easily by natural forces to produce all of the variety we see among species today, why does any line exist at all that is stable enough and distinct enough to be called a species? Why is the world not filled with intermediate forms of every conceivable kind? In fact, the world corresponds much more closely to what can be expected from the intelligent design point of view: it is filled with distinct and stable species that retain their identity over long periods of time, and intermediate forms expected by Darwinists are missing.

This leads to the second problem for the Darwinists, the problem of stability. For example, why has an organism like the shark not changed for 150 million years (by the conventional time scale)? W. H. Thorpe, Director of the Subdepartment of Animal Behavior at Cambridge University in England said:

> What is it that holds so many groups

of animals to an astonishingly constant form over millions of years? This seems to me the problem now (for evolution)—the problem of constancy, rather than that of change.[6]

Third, Darwinists have failed to set forth a convincing explanation of how new functional information is introduced into the genotypes of the vast variety of fossil and living organisms.

Suggested Reading/Resources

The Nature and Origin of the Biological World, E.J. Ambrose, New York: Halsted Press, 1982. An excellent overview of cell biology with emphasis on information theory and the role of intelligence.

Darwin on Trial, Phillip E. Johnson, Washington, D.C.: Regnery Gateway, 1991. A legal authority evaluates the strengths and methodologies of Darwinist arguments.

References

1. T. Dobzhansky, 1958. "Species after Darwin," in *A Century of Darwin.* ed. Samuel A. Barnett, London: William Heinemann, p. 48.
2. Charles Darwin, 1859. *The Origin of Species.* 1968 Reprint, London: Penguin Books, p. 292.
3. E. Mayr, 1963. *Animal Species and Evolution.* Cambridge, Mass: Harvard University Press, p. 12.
4. G. G. Simpson, 1964. *This View of Life.* New York: Harcourt, Brace & World, p. 81.
5. E. Sober, 1984. *The Nature of Selection.* Cambridge, Mass: MIT Press, p. 197; O. Rieppel, 1990. *J. Hist. Biol.* 23, 291-320 (see esp. p. 303); B. C. Goodwin, 1990. *J. Science Progress,* Oxford 74, 227-244; J. E. McDonald, 1983. *Annl Rvws of Ecol and Systematics* 14, 77-102; C. Mann, 1991. *Science* 252, 378-381 (esp. p. 379); E. J. Ambrose, 1982. *The Nature and Origin of The Biological World.* New York: Wiley, Halsted, pp. 119-131 (esp. p. 131).
6. W. H. Thorpe as quoted in G. R. Taylor, *The Great Evolution Mystery.* New York: Harper & Row, pp. 141-142.

EXCURSION CHAPTER 4

The Fossil Record

Introduction

Science has long been celebrated as immune to the subjectivity which affects other areas of man's understanding. It has been widely believed that the methodology of science provides a sort of filter to remove the distortions of knowledge that could come from the individual scientist's philosophy or values. This concept has dominated because most scientific theories, such as the germ theory of disease, the theory of gravitation, and Mendel's theory of heredity, are theories about how things operate. So, most science theories can be checked simply by comparing them with what actually occurs. If we have a theory about the earth orbiting the sun, for example, we could propose to test it by predicting a solar eclipse, and then observing to see if it occurs as predicted. Such an empirical check is why some scientists consider science to be value-neutral. Regardless of their individual philosophies or points of view, several scientists conducting the experiment the same way will get the same results. Most scientific theories describe these kinds of repeatable phenomena.

But Darwin's theory that all living things evolved by natural selection is very different from most other scientific theories. It is a theory about unique past events, events that have come and gone. However life originated in the first place, by intelligent design or spontaneous generation, or however the giraffe or the aardvark originated, they are not "re-originating." These are one-time events. Even if these events are part of a natural law process, they are unique and nonrepeating. The famous geneticist Theodosius Dobzhansky said:

> . . . evolutionary happenings are unique, unrepeatable, and irreversible. It is as impossible [experimentally] to turn a land vertebrate into a fish as it is to effect the reverse transformation.[1]

A biological origin by intelligent design would also be unique, unrepeatable, and irreversible. So theories of origins can't be tested by direct empirical test like the theories mentioned earlier. This fact leaves origins theories open to subjectivity and to the interpretive elements of individual viewpoints and values. Since origins are one-time events and origins theories cannot be checked against recurring phenomena, they must be checked by other means. This is done by evaluating the plausibility of origins theories in two ways. First, we have to ask if there is convincing similarity between present and past causes. That is what Darwin did with the artificial selection used by breeders to get improved stock. He said it was analogous to natural selection the cause in the past for species transformation. When examples of natural selection in the present were also presented, it merely strengthened the case.

The second way to check the plausibility of an origins theory is to consider circumstantial evidence. In a murder case with no eyewitnesses a lawyer must rely on the strength of circumstantial evidence. For example, Smith is accused of shooting Jones with a .38 revolver. Smith's fingerprints are found on a .38 revolver beside Jones' body. Smith had left little doubt about his dislike for Jones the night before the murder. The case built on circumstantial evidence is not proof, though it may sound plausible and incriminating. Even so, the jury's belief that Smith is guilty may be more a product of their subjective feelings than they realize. This is why Smith is entitled to defense counsel, so that both sides of the issue can be considered. A case of this sort often turns on the performance of the defense attorney, whose skill in marshalling the facts is critical. Can the plausibility of the case against the defendant be shaken or even overturned and replaced with another scenario showing innocence? Only if the defense attorney skillfully uses the circumstantial evidence.

Major Features of the Fossil Record

In the absence of eyewitness testimony the fossil record provides circumstantial evidence to paleontologists and biologists. There are three notable features of the fossil record that must be considered in attempting to find out how life began and came to exist in its profusion of forms.

1. The vast majority of the known animal phyla (over 95%) are either known or believed to have appeared within a geologically "brief" period (estimates range from 10 to 40 million years). Thereafter, new phyla stop appearing throughout the geological record. The phyla are the major groups of life forms, based upon large differences in morphology, especially basic body plans.

2. After fossils first appear in the record they persist largely unchanged through many strata (a phenomenon called **stasis**) (STAY-sis), then frequently they suddenly disappear from the record.

3. Fossil species are fully formed and functional when they first appear in the record. There is a conspicuous lack of evidence for graded series of in-between fossils. Instead, numerous gaps exist throughout the fossil record.

An additional issue concerns the matter of the earth's age. While design proponents are in agreement on these significant observations about the fossil record, they are divided on the issue of the earth's age. Some take the view that the earth's history can be compressed into a framework of thousands of years, while others adhere to the standard old earth chronology. In this chapter, we will examine the three features outlined above.

Early Expectations of the Fossil Record

Darwin used the fossil record as circumstantial evidence that the species had originated through natural means. Of course, DNA was unknown in Darwin's day. Ever since that time many people have cited the fossil record as the best evidence for evolution. Why? Because it could potentially provide impressive circumstantial evidence. If all living things are related to each other through evolutionary descent as Darwin said, and as is often pictured in a typical evolutionary tree (see Figure 4-1), would we not expect to find many intermediate or **transitional** forms between major **taxa**? In fact, the major groups should blend into one another with "evolutionary trails" of innumerable transitional forms connecting the fossil organisms found. The observations of such transitional forms would be excellent material out of which to build a theory of evolution. At the base of the tree would be the first form of life, ancestor to all that follows. Over time, the tree grows and branches out as new species appear. As more species evolve, clusters of genera gradually develop, forming families, and eventually orders; classes and phyla are formed in much the same way. As the branching continues, ultimately all of the phyla would emerge, some near the base, but many high in the branches. Yet Darwin was aware of no such fossil evidence. Scientists simply had not discovered among the fossils the "missing links" that should exist were his theory correct. In *The Origin of Species* he asks rhetorically:

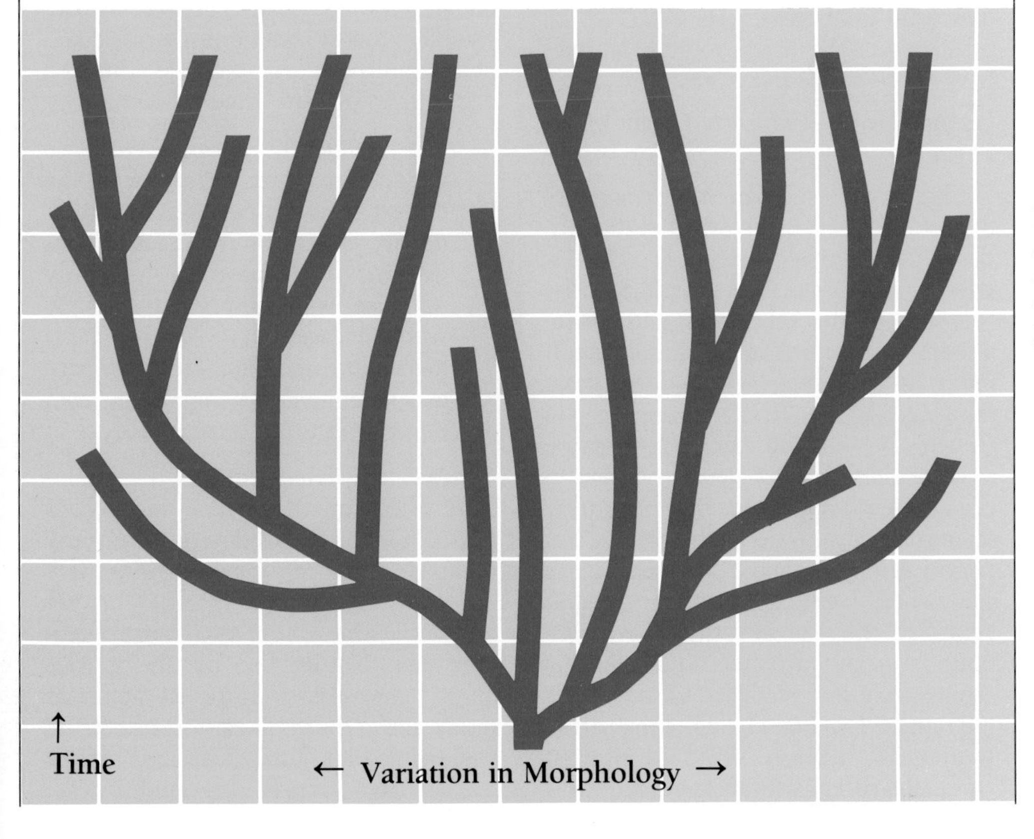

Figure 4-1. *A generalized phylogenetic tree, showing the customary gradual pattern of evolution.*

> . . .The number of intermediate varieties, which have formerly existed on the earth, [must] be truly enormous. Why then is not every geological formation and every stratum full of such intermediate links? Geology assuredly does not reveal any such finely graduated organic chain; and this, perhaps, is the most obvious and gravest objection which can be urged against my theory.[2]

To his credit, Darwin acknowledged that this was a serious problem. His honesty was disarming and many scientists were sympathetic to the ideas he had developed. After all, they thought, maybe some of the transitional fossils would later be found. Darwin himself expected that these transitional forms would turn up as scientists searched. Today, over 125 years after the publication of Darwin's theory, we know of thousands of fossil organisms that were unknown to Darwin. But the gaps between the major groups of animals have not been filled.

Feature No. 1–*The Early Origin of the Phyla*

Easily the most dramatic feature of the history of life is the origin of the fundamental groups—the phyla. This is where the greatest differences exist (between major body plans) and consequently where the most modification must have taken place. Yet the great majority of the living animal phyla (roughly 30—the number varies because scientists disagree on details of how to classify them) appear (or are thought by scientists to appear) in a remarkably brief period of time, geologically speaking, of somewhere between 10 and 30 million years at the Precambrian-Cambrian boundary, and are not connected by evolutionary intermediates. (If one takes the standard dating scheme of the earth's strata, 30 million years is almost "momentary.") It is suspected that if additional phyla are found, they too, will trace back to this same period (see Figure 4-2).

Only two exceptions to this striking feature are known to present day paleontology. First, no fossil record of the flat worms (phylum Platyhelminthes) is known until later, yet most scientists believe they must have existed before most of the other phyla. Second, another dozen "phyla" (a classification for these organisms that is disputed) of now extinct soft-bodied animals have been found in the Middle Cambrian, but this stretches the 30 million years only to 40, and the true origins of these extinct groups might date back to the same period as the other approximately 40-42 phyla. Figure 4-2 helps to visualize how broad this feature is, characterizing as it does nearly all known animal phyla. University of California at Santa Barbara paleontologist James W. Valentine has speculated that some phyla are still undiscovered and estimates that at least 60 phyla originated at this enormous beginning. He remarked about this surprising feature:

> We consider these estimates to be very conservative. If they are on the correct general order, then the evolutionary events near the Precambrian-Cambrian boundary not only occurred with unexpected rapidity within lineages but involved so many higher taxa as to form an evolutionary explosion without precedent, both rapid in pace and broad in scope.[3]

As if that weren't enough, the "explosion" of new animal phyla in the early to Middle Cambrian is followed by a nearly complete "silence"; new phyla simply stop appearing throughout the remaining 500 million years or more of geologic time! Not only so, but the appearance of new classes drops off almost as dramatically. For instance, over half the 62 well-defined and easily fossilized marine classes have already appeared by the

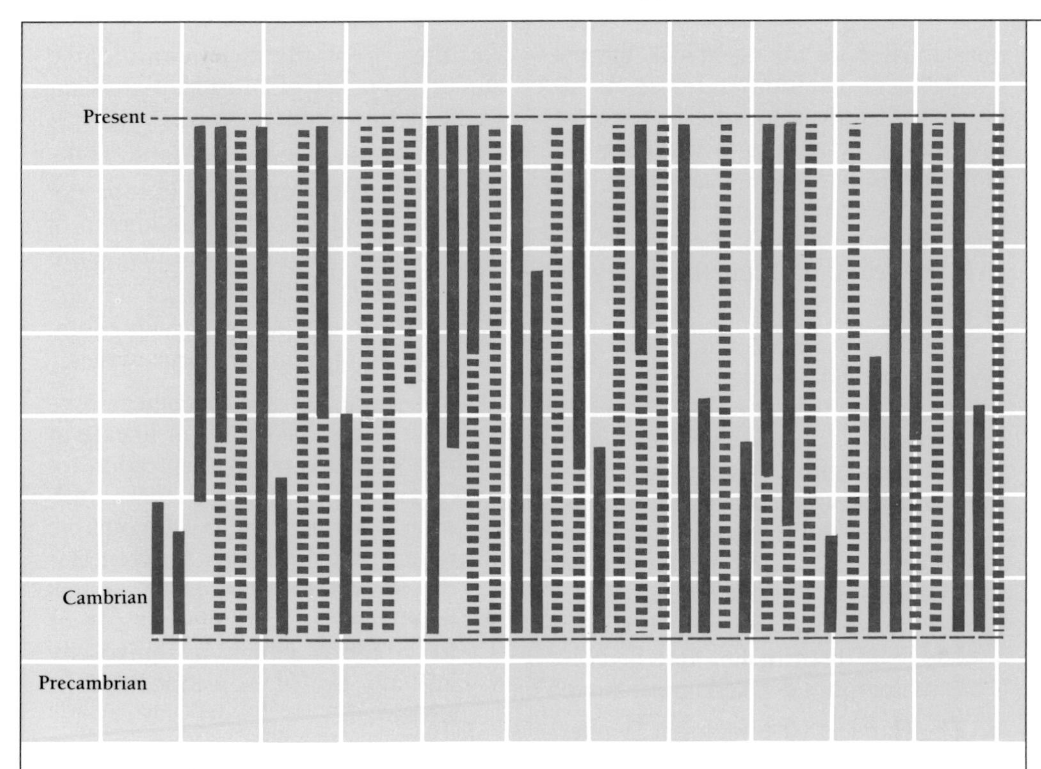

Figure 4-2. *A generalized schematic of the fossil record, designed to show the Cambrian origins of nearly all animal phyla in relation to the overall time scale of the history of animal organisms. Dotted lines represent the presumed existence of phyla, not the fossil record.*

close of the Cambrian, and another 29% first occur in the Ordovician.[4]

According to Valentine and his coauthors, the number of new orders also declines over time:

> Appearances of marine orders are somewhat more dispersed but again show a pattern of major innovation during the early Paleozoic.[5]

This hardly matches the expectations arising from the evolutionary tree depicted earlier.

This nearly simultaneous appearance of most known phyla is more remarkable when we consider that the variation within a phylum is quite small compared to how much the phyla vary from one another. In other words, there is more morphological distance between two phyla than separates representatives within the phyla themselves. This means that the origins of new phyla are evolution's greatest achievements in diversifying life forms. Yet, crowded as these achievements were into the first 5% of the fossil record for animals, there is an unexpected lack of fossils bridging the evolutionary distance between the phyla to document evolutionary origins for them. Although the extremely early and isolated appearances of most of the phyla are certainly a dramatic feature of the fossil record, this pattern receives little attention in most biology textbooks.

Features No. 2 and 3—*Fossil Stasis and Gaps within the Phyla*

This same characteristic lack of transitions is found in much of the rest of the fossil record. Even if we pool fossil data from all over the world and from rocks regardless of their proposed ages, we cannot form a smooth, unambiguous transitional series linking, let's say, the first small horse to today's horse, land-dwelling mammals to whales, fishes to amphibians, or reptiles to mammals. As more fossils have been discovered, the gaps have become more pronounced. David Raup, Curator of Geology at the Field Museum of Natural History in Chicago says:

> Well, we are now about 120 years after Darwin, and knowledge of the fossil record has been greatly expanded . . . ironically, we have even fewer examples of evolutionary transition than we had in Darwin's time. By this I mean that some of the classic cases of Darwinian change in the fossil record, such as the evolution of the horse in North America, have had to be discarded or modified as a result of more detailed information.[6]

The Harvard paleontologist Stephen Jay Gould says:

> The extreme rarity of transitional forms in the fossil record persists as the trade secret of paleontology. The evolutionary trees that adorn our textbooks have data only at the tips and nodes of their branches; the rest is inference, however reasonable, not the evidence of fossils.[7]

He goes on to identify two features in the history of most fossil species:

> 1) Stasis—most species exhibit no directional change during their tenure on earth. They appear in the fossil record looking much the same as when they disappear; morphological change is usually limited and directionless;
>
> 2) Sudden appearance—in any local area, a species does not arise gradually by the steady transformation of its ancestor; it appears all at once and fully formed.[8]

Another prominent paleontologist, Steven M. Stanley of Johns Hopkins University, writes about the genus level:

> . . . despite the detailed study of the Pleistocene mammals of Europe, not a single valid example is known of phyletic [gradual] transition from one genus to another.[9]

With respect to the plant kingdom, Harold C. Bold, a morphologist, writes:

> It is, however, when we come to consider the actual course or lineage in the subsequent diversification of organisms, . . . that we meet with disappointment and frustration if we rigorously distinguish between evidence and speculation. . . . At this time there are no known living or fossil forms which unequivocally link any two of the proposed divisions.[10]

What can we conclude? The extreme rarity of fossil transitional forms between the various types of plants, and between the various types of animals, is a vexing problem for Darwinian thought. What is the meaning of these gaps? How should they be interpreted?

The Meaning of Gaps in the Fossil Record

Several interpretations have been offered to resolve this problem:

1. Imperfect Record

The gaps result from the imperfect nature of the fossil record, only a small part of which was preserved, and it seems unlikely that future research will fill them. Support for the theory of evolution must come from other fields of study.

2. Incomplete Search

The gaps result from both the imperfection of the fossil record and the incomplete sampling of that record.

Future research will be able to fill in some of the gaps.

3. Jerky Process

The gaps result from the nature of the Darwinian process more than from the nature of the fossil record or amount of research done so far. In other words, Darwin was incorrect when he taught that transformation was smooth and gradual. According to this school of thought, we should expect gaps in the fossil record because living things emerge in sudden jumps in which there were at best only a few transitional forms preserved. Darwin's view is represented by the type of tree of life shown earlier in Figure 4-1. The sudden, jerky (punctuated) type of evolution is pictured in Figure 4-3. The short, connecting dotted lines represent proposed rapid Darwinian changes in population too small to be recorded in the fossil record. This more recent view is called **punctuated equilibrium.** This explanation places more emphasis on stasis—the fact that species are relatively stable during long periods of time. In Chapter 3 we saw some of the biological reasons for stasis. This stability (a record of biological "equilibrium," according to the theory) is occasionally punctuated by bursts of Darwinian change in small populations of species. Several new species may be formed quite rapidly over the course of just a few thousand years.

However, because punctuated equilibrium is more random in its operation than the traditional view of Darwinism (**gradualism**), it may take several trial species or steps of directional change

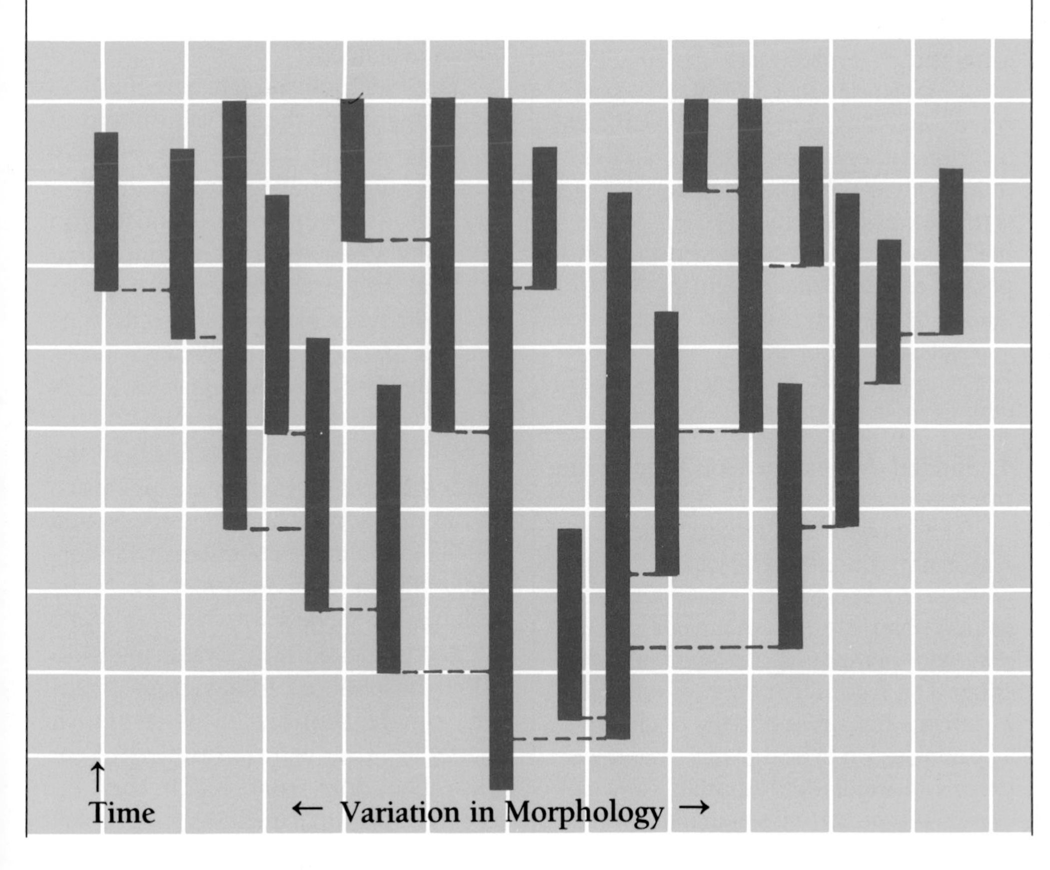

Figure 4-3. *A generalized phylogenetic tree, representing a "punctuated" pattern of evolution. The dotted lines represent unobserved speciation events.*

(represented by the short horizontal dotted lines in Figure 4-3) to establish a new species which will stabilize and develop into a population. Yet these transitional species leave little fossil trace because of their relatively small numbers. (Even among large populations, fossilization is highly unusual, requiring specialized conditions.) According to the theory, most of these "trial species," formed more or less randomly from the parent species soon become extinct. One or two new species, significantly different from the parent species, survive and multiply.

But notice that this theory must rest on the gaps in the fossil record, an absence of data. In fact, there is scant positive evidence for the theory's key proposal—the rapid spurts of evolution. As a broad idea, it may be that punctuated equilibrium predicts the observed pattern of fossil distribution, that is, species stasis separated by evolutionary space. But we must remember that fossils are assemblages that originated from remarkable functional information. It seems fair to say that the theory of punctuated equilibrium is not compelling as long as there remains neither fossil record of the genetic transport of this information (in the form of graded series of transitional forms), nor an explanation convincing even to many Darwinists, for how it was delivered to the various fossil organisms.

4. Sudden Appearance or Face Value Interpretation

The known fossil record is assumed reasonably complete. The gaps show that while some species may have arisen by gradual change, at least the major taxa did not, and perhaps many species didn't either. The fossil record shows that most organisms remain essentially unchanged. The conclusion to be drawn is that major groups of plants and animals have coexisted on the earth independent of each other in their origins, which must be explained in some way other than Darwinian evolution.

Scientists should not accept the face value interpretation of the fossil record without also exploring the other possibilities, and even then, only if the evidence continues to support it. The imperfect record and incomplete research interpretations above are attempts to make the fossil record compatible with the Darwinian view of origins, which teaches step-wise evolution from one form of life to another. Both of these views acknowledge that the present existence of gaps in the fossil record is not in agreement with what is expected by Darwinian theory. The question many scientists are asking is, How long should we continue to entertain these possibilities in the absence of evidence? Should other possibilities be ignored?

The intelligent design hypothesis is in agreement with the face value interpretation and accepts the gaps as a generally true reflection of biology and natural history. A growing number of scientists who study the fossil record are concluding that the structural differences between the major types of organisms reflect life as it was for that era. This view proposes that only the long-held expectations of Darwinian theory cause us to refer to the in-between areas as gaps. If this is so, the major different types of living organisms do not have a common ancestry. Such a conclusion is more consistent with currently known fossil data than any of the evolutionary models.

According to the face value interpretation, the fossil record of various organisms can be represented as shown in Figure 4-4. The vertical lines represent taxa. The lines that reach the top represent taxa that include species still

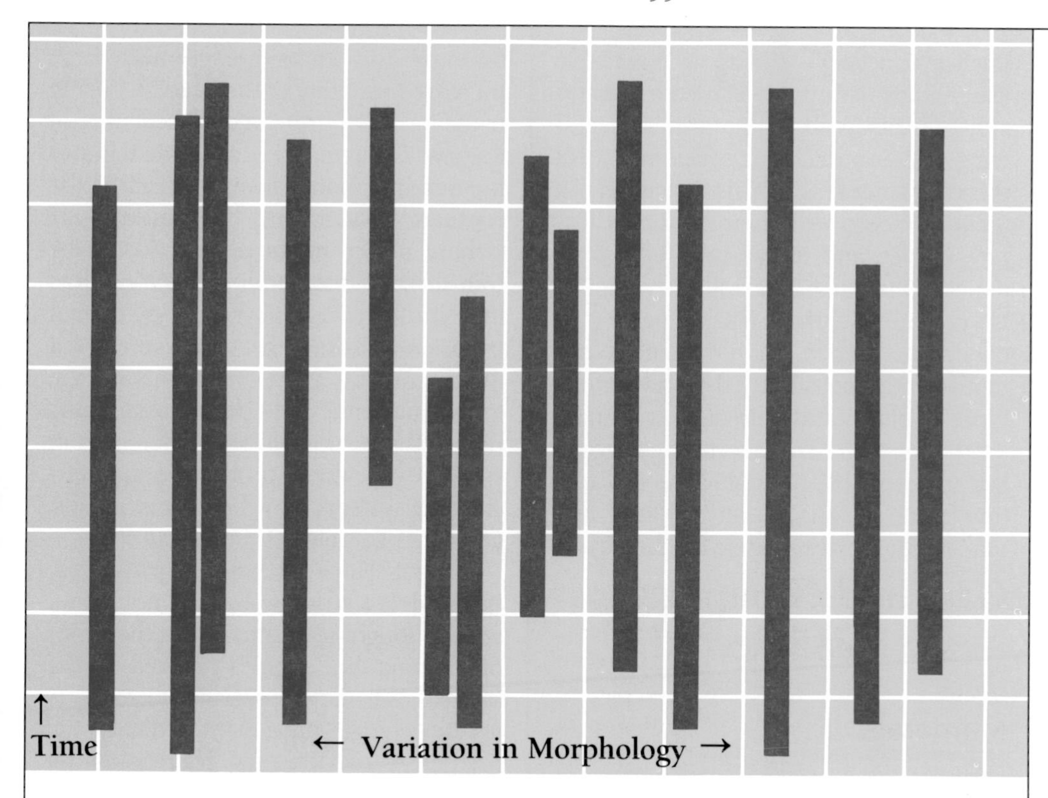

Figure 4-4. *The pattern of phylogenetic origins, according to the face value interpretation of the fossil record.*

alive today. Those that end before reaching the top represent extinct forms like *Archaeopteryx* (ar-kee-OP-tuh-rix) or the dinosaurs. The vertical lines do not converge at the bottom as they did in Figure 4-1. This indicates that as more fossils are being discovered, organisms within distinct groups, not transitional forms, are being found in older and older strata.

The intelligent design view is also consistent with Feature No. 2—the observation that some forms of life have undergone little or no change (by conventional reckoning) for tens or hundreds of millions of years. Some of these forms are still surviving today. Table 4-1 lists examples of some of these "living fossils."

Darwinists object to the view of intelligent design because it does not give a natural cause explanation of how the various forms of life started in the first place. Intelligent design means that various forms of life began abruptly through an

Examples of "Living Fossils"

Organisms	First Appearance	Years Ago
Aardvarks	Early Miocene	20 My*
Alligators	Early Oligocene	35 My
New World Porcupines	Early Oligocene	35 My
Snapping turtles	Late Paleocene	57 My
Sirens (amphibians)	Late Cretaceous	80 My
Sturgeons	Late Cretaceous	80 My
Echinoneid sea urchins	Late Cretaceous	80 My
Bowfin fishes	Middle Cretaceous	105 My
Galatheid crabs	Middle Jurassic	170 My
Horseshoe crabs	Early Triassic	230 My
Notostracan crustaceans	Late Carboniferous	305 My
Kakabekia (possible Precambrian protozoan)		1,900 My

*My = millions of years

Table 4-1. *A sample of organisms remaining largely unchanged over vast periods of time.*

intelligent agency, with their distinctive features already intact—fish with fins and scales, birds with feathers, beaks, and wings, etc. Some scientists have arrived at this view since fossil forms first appear in the rock record with their distinctive features intact, and apparently fully functional, rather than gradually developing. No creatures with a partial wing or partial eye are known. Should we close our minds to the possibility that the various types of plants and animals were intelligently designed? This alternative suggests that a reasonable natural cause explanation for origins may never be found, and that intelligent design best fits the data.

Gaps and Groupings in the Fossil Record

Mammals

A most impressive example of transition to which Darwinists point is the series bridging from the reptiles to the mammals. This class-level transition is to have taken place through a group of mammal-like reptiles called Therapsids (thuh-RAP-sidz). Among the several Therapsid lineages were the dominant land-dwelling vertebrates from the middle of the Permian period to the middle Triassic. Indeed, it does appear that they provide Darwinists with a superior example of a transitional series. In 1987, evolutionary biologist James Hopson published an article describing a series of eight Therapsid skulls that made up a fairly well-filled-in sequence of intermediate types, apparently leading to the ninth, an early mammal named *Morganucodon* (mor-guh-NOO-kuh-don).[11]

Hopson detailed several characters exhibited by the series, characters that progress together toward the mammalian body plan. These include: 1. Change in the way the limbs are connected. 2. Increased mobility of the head. 3. Fusing of the palate. 4. Improved musculature of the jaw. 5. Migration of the articular and the quadrate bones from the back of the reptile's jaws toward the middle ear (where in the mammal they would be transformed into auditory ossicles). It is the simultaneous movement of several traits, says Hopson, that clearly infers that the Therapsids are a continuous lineage to the mammal. (Of course, fossils can't record the potentially vast differences in systems like the reproductive and circulatory systems, nor the organs, glands, and other soft tissues they entail.)

What Hopson actually presented, however, is a structural series, not a lineage. Although he predicts "that the series of mammal-like reptiles ordered on the basis of morphology will also form a series in geologic time,"[12] in actuality, the first three of Hopson's Therapsids are contemporaries from two separate orders, and some are not thought to be mammalian ancestors. Rather than older, the fourth is more recent than the fifth, and the final Therapsid is more recent than the mammal (Morganucodon) presented as its descendent!

It is legitimate to assemble a morphological series for the purpose of speculating about which skull is structurally intermediate to which others, but it is certainly not in the interest of education if it is presented as a single path of descent—an actual evolutionary lineage.

There are numerous fossil Therapsid species in the record. In fact, Douglas Futuyma said:

> The gradual transition from Therapsid reptiles to mammals is so abundantly documented by scores of species in every stage of transition that it is impossible to tell which Therapsid species were the actual ancestors of modern mammals.[13]

Without doubt, the Therapsids are highly suggestive of a Darwinian lineage. But the fact that a number of parallel Therapsid lineages approach the threshold of the mammalian class raises two additional questions.

First, if mammals arose from just one of these lineages, then the others are not ancestral to them. But if several unrelated species have the same mammal-like features as the "actual ancestor," *how compelling are these features as evidence of ancestry for the skeptical inquirer?* Do they *require* the Darwinian interpretation, or merely *suggest* it?

On the other hand, if mammals arose several times (as suggested by many researchers) from several different Therapsid species, then we must accept that accidental mutations crafted the extraordinary, precisely integrated parts of the mammalian ear. Moreover, they did this many times, each independently, a claim that seems severely strained.

We might speculate that the information for these intricate devices may have existed, unexpressed, in the genome of the earliest Therapsids before they diversified. Then it could have been passed on, as part of the genetic blueprint, to the several Therapsid variants that later arose. But here is the problem. Had natural selection been involved, the crafting process could not have been "hidden" in unexpressed genetic material; natural selection only acts on expressed traits or structures. So the evidence seems to support the existence of a common blueprint not developed by descent.

The absence of unambiguous transitional fossils is illustrated by the fossil record of whales. The earliest forms of whales occur in rocks of Eocene age, dated some 50 million years ago, but little is known of their possible ancestors. By and large, Darwinists believe that whales evolved from a land mammal. The problem is that there are no clear trans-

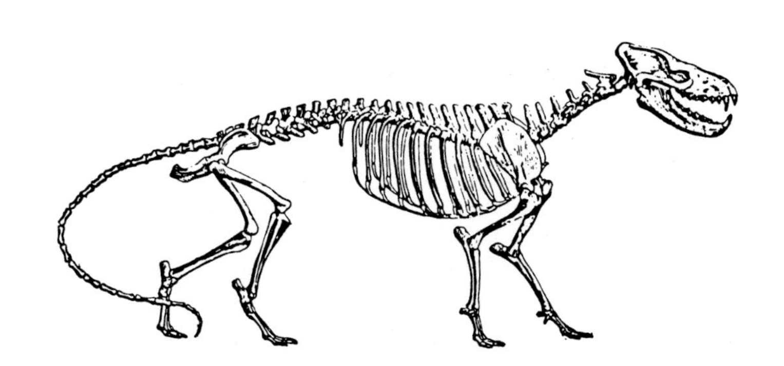

Figure 4-5. ***Skelelton of Basilosaurus, an Eocene whale (upper) compared with the skeleton of Mesonyx (lower). The Mesonyx has been suggested as the terrestrial ancestor of the whale.***

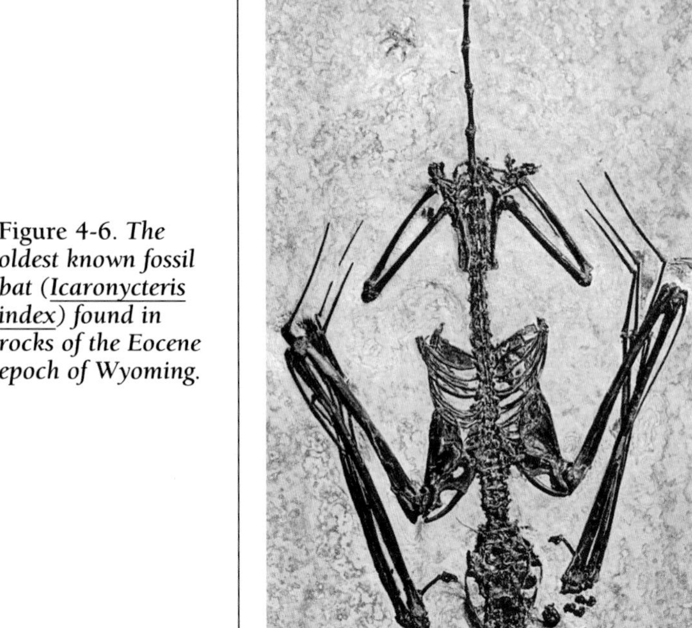

Figure 4-6. *The oldest known fossil bat (Icaronycteris index) found in rocks of the Eocene epoch of Wyoming.*

itional fossils linking land mammals to whales. If whales did have land-dwelling ancestors, it is reasonable to expect to find some transitional fossils. Why? Because the anatomical differences between the two are so great that innumerable in-between stages must have paddled and swam the ancient seas.

Figure 4-5 shows the skeleton of *Basilosaurus* (BA-sil-ah-SOR-us) an Eocene whale, together with the skeleton of *Mesonyx*, a carnivore suggested by some Darwinists as representing its ancestor.

Recently, small pelvic limb and foot bones were found "in direct association" with a fossil of Basilosaurus in an Egyptian desert.[14] These have been touted as evidence of the land-dwellers-to-whales theory. While this find is interesting, students should be told that portions of the same structures were found with one of the first Basilosaurus finds in the late 19th Century. What is different now, is that enough of these structures have been recovered to suggest that they functioned as guides to mating, and were not vestigial as originally thought. (For the design proponent, who rejects the blind watchmaker hypothesis, function is a great help in explaining the existence of a structure.)

Compare the skulls, backbones, ribs and appendages in Figure 4-5. Great structural differences between the two forms are obvious. Even though the Basilosaurus limb and foot bones are intermediate, they are hardly a substitute for the important line of missing transitional forms.

Moreover, Stephen Gould has calculated the time allowed by the fossil occurrences as far too brief; the changes required to evolve a Basilosaurus from the Mesonyx by punctuated speciation events are more than two orders of magnitude too great, even from the Darwinian perspective. If we are to accept the existence of transitional species leading to the whale, say design proponents, we must do so without clear evidence.

The case of bats (order Chiroptera, kye-RAHP-tuh-ruh) is similar to that of whales, although the fossil record for bats is more fragmentary. Fossils consist mainly of teeth and skull fragments. An example of the oldest known fossil bats was found in early Eocene rocks in Wyoming. Its skeleton is shown in Figure 4-6. There is no fossil record of the evolution of bats. Still, many scientists believe that it must have happened.

The whole mammal class, in fact, has been classified into as many as 32 different orders, each represented in the fossil record (see Figure 4-7). But a well-known riddle has become associated with these mammalian orders. Their isolation in the fossil record is striking. Scientists would like to know, What were their ancient ancestors?

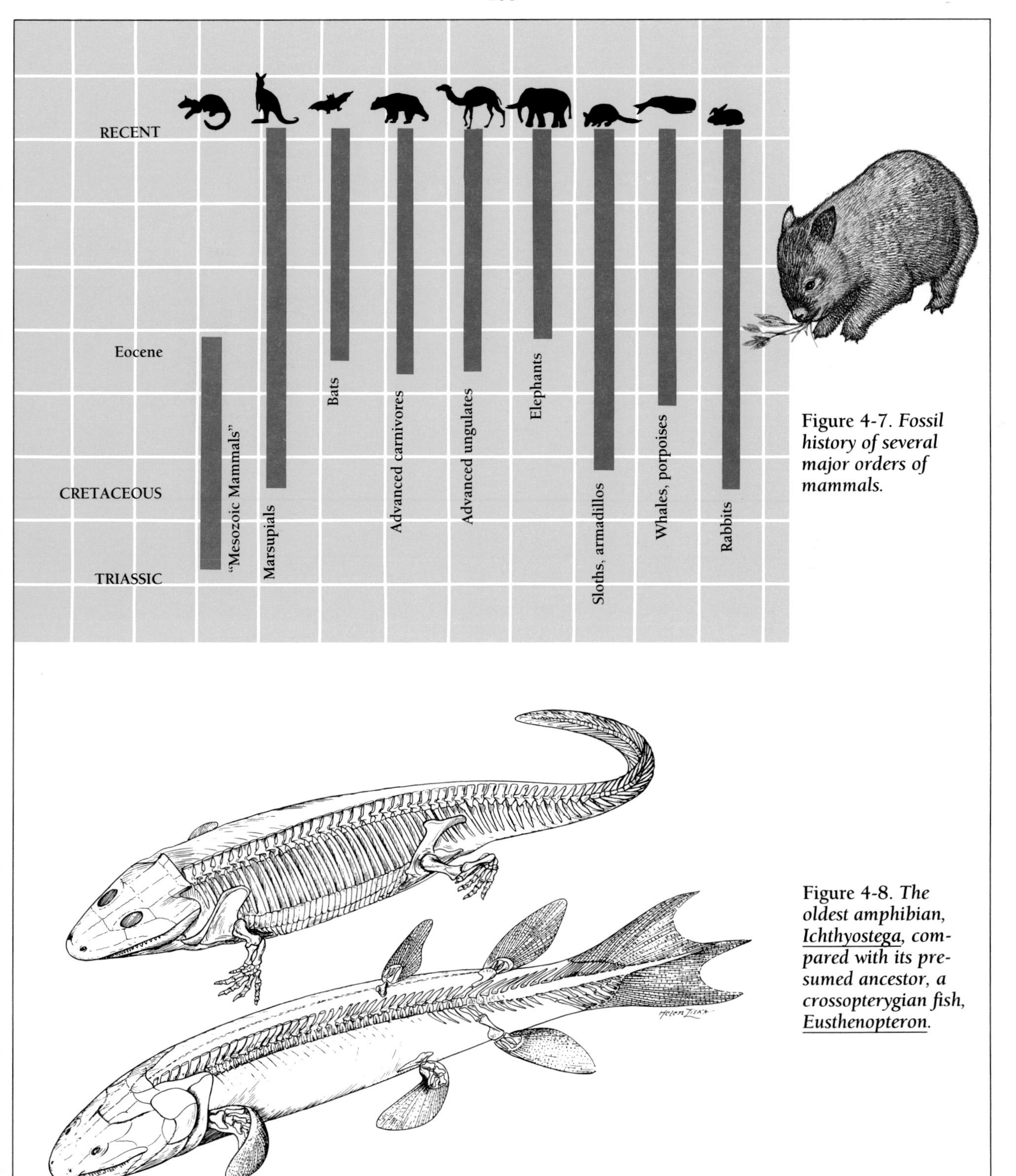

Figure 4-7. *Fossil history of several major orders of mammals.*

Figure 4-8. *The oldest amphibian, Ichthyostega, compared with its presumed ancestor, a crossopterygian fish, Eusthenopteron.*

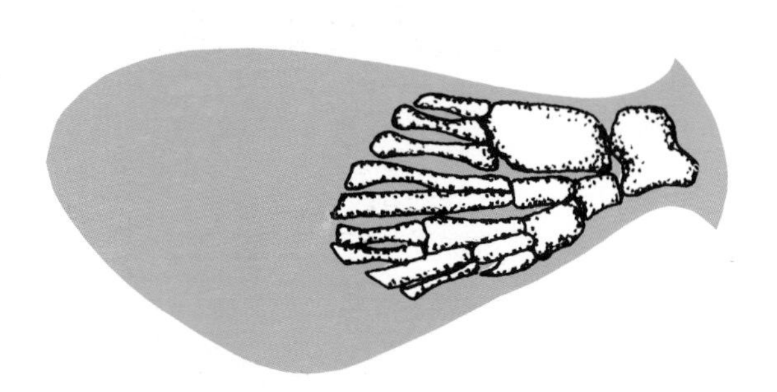

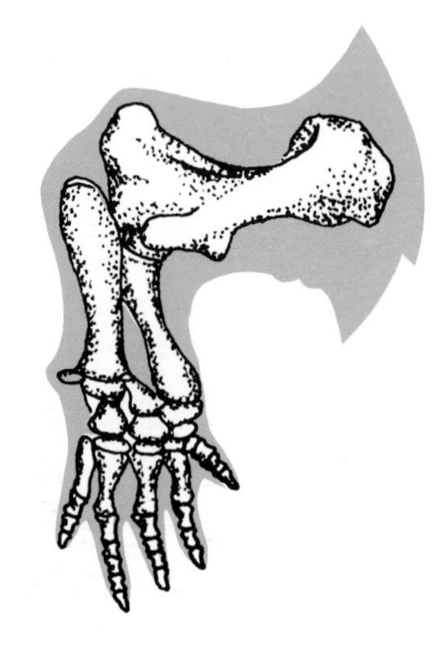

Figure 4-9. *The bones of a crossopterygian fin compared to the bones of an ichthyostegid forelimb.*

Amphibians

Let us turn from mammals to amphibians. Darwinists believe that the first amphibians (the *labyrinthodonts,* la-buh-RIN-thuh-dontz) evolved from early fish known as *crossopterygians* (KRAW-SAHP-tuh-RIJ-nz) or lobe-finned fish. A very similar lobe-finned fish swims the Indian Ocean today. Look at the comparison of the oldest known amphibian skeleton, *Ichthyostega* (IK-the-o-STAY-ga), with a crossopterygian fish shown in Figure 4-8.

If crossopterygians really did evolve into amphibians, tremendous changes must have taken place. Fins must have been transformed into forelimbs (see Figure 4-8). The skull had to change from two parts to a single, solid piece. The hip bones had to enlarge and become attached to the backbone. Numerous changes must also have occurred in organs, muscles and other soft tissues. For example, the air bladder of the fish had to be transformed into the lungs of the amphibian.

Though just a few of the many examples possible, these are enough to show how large the differences between early fish and amphibians really were. How many different transitional species were required to bridge the gap between them; hundreds? Even thousands? We don't know, but we do know that no such transitional species have been recovered. Moreover, we have no fossil evidence of the evolution of the crossopterygians from other fish. Two large gaps thus exist in the fossil record between ordinary Devonian fish (325 million years ago) and amphibians; one between ordinary fish and crossopterygians, and an even larger gap between these lobe-finned fish and amphibians.

Archaeopteryx

It is interesting that nearly every organism possesses the defining characteristics of its taxon. In fact, only a handful of transitional forms have been proposed.

Figure 4-10

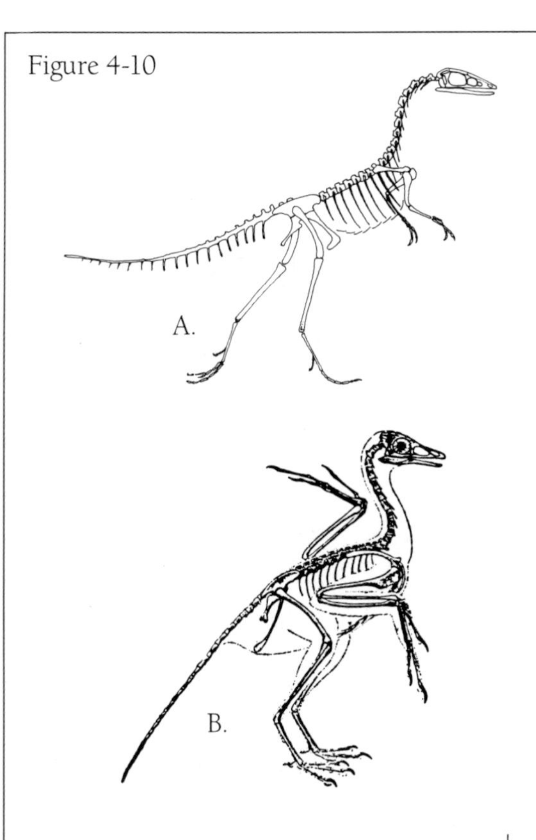

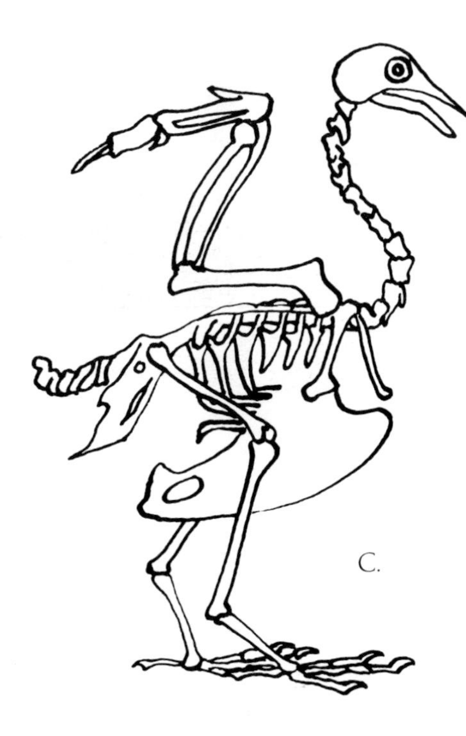

Figure 4-10. *Skeletons of a Therapod dinosaur (A) representing the presumed reptile ancestor of birds, the Jurassic fossil, Archaeopteryx (B) and a modern bird (C).*

Let's look at the most famous of all, *Archaeopteryx*. The oldest fossil evidence suggestive of birds are the remains of *Archaeopteryx*, six examples of which have been found from the Jurassic period. Darwinists believe that *Archaeopteryx* evolved from a therapod dinosaur like Compsognathus (KOMP-suhg-NA-thuss). See Figure 4-10.

Archaeopteryx had teeth like many (but not all) reptiles and like some later fossil birds such as *Hesperornis* of the Cretaceous Period.

As we saw in the section about mammals, the simultaneous movement of several features in the Therapsids toward the mammalian body plan was taken by Darwinists as evidence that they formed a continuous evolutionary lineage. Presumably these features were fashioned under a "selective regime" favoring the mammalian lifestyle. Thus they would all inch toward that lifestyle together, as the genome was replenished by continuous mutations from which to select.

Consider how this principle applies to *Archaeopteryx*. The name *Archaeopteryx* means "old wing." Yet ironically, the wings, or at least the feathers, are the most modern part of this creature. If we judged strictly by its feathers, we would be convinced that *Archaeopteryx* was an accomplished flyer, because they provided an airfoil superbly adapted for flight. But birds have an adaptational package of several characteristics—known as the "avian complex"—that suits them for flight. This includes such necessities as a sternum and large wishbone (furcula) providing adequate surface for the attachment of pectoral muscles, the anchoring of major wing feathers to the ulna by ligaments, and several others.

Figure 4-11. ***A schematic representation of the fossil record of Archaeopteryx. The graduated rectangles represent an untold number of missing intermediates between reptile and bird.***

In place of these and other avian structures, however, *Archaeoteryx* has typical reptilian characteristics:[15] 1. Gastral ribs instead of a sternum. 2. A small furcula. 3. A vertically positioned femur. 4. A saurian pelvis. 5. A manus with three clawed fingers. 6. A long spinal tail. 7. Ribs without uncinate processes. 8. Reptilian growth pattern (continuous). 9. Evidence of reptilian type lung.

Clearly, the characteristics of *Archaeopteryx* are not predicted by Darwinism for a transition between reptiles and birds. In Darwinian theory, only under a regime of selective pressure for flight would fully modern feathers emerge. David Wilcox, Professor of Biology at Eastern College, points out that there should, therefore, be a progressive acquistion of the flight features running parallel to the plumage of *Archaeopteryx*; however, they are simply nowhere to be found.[16] In their place are the reptilian features setting *Archaeopteryx* apart from birds. Moreover, once *Archaeopteryx* acquired feathers, the selection pressure for the avian complex would intensify further. This strongly suggests that something other than mutation and natural selection is responsible for this exotic fossil.

John Ostrom of Yale University believes that *Archaeopteryx* was incapable of flight and used its wings to chase down insects instead. Other scientists, however, point to the primary flight feathers as evidence of flight. If only we could find a fossil showing scales developing the properties of feathers, or lungs that were intermediate between the very different reptilian and avian lungs, then we would have more to go on. But the fossil record gives no evidence for such changes (see Figure 4-11).

How should this strange fossil be interpreted? The fact that it possessed reptilian features not found in most birds does not require a relationship between birds and reptiles. A fossil can be intermediate (a matter of morphology) without being transitional (a matter of origin). It is transitional only if it is part of a lineage—one of a series of generations in which in-between stages led gradually from one group to another. Some authorities feel that *Archaeopteryx* was a side branch that did not lead to birds.[17] Similarly, some design proponents believe *Archaeopteryx* fails to fill the gap between reptiles and birds because it is intermediate (and that in only one or two features) not transitional. Acknowledging that the transition problem applies to *Archaeopteryx*, Colin Patterson, Curator of fossils at the British Museum where the leading *Archeopteryx* fossil is kept, said:

> I will lay it on the line—there is not one such fossil [a fossil that is ancestral or transitional] for which one could make a watertight argument.[18]

The Origin of Man

Human-like fossils are more interesting to most of us than any other kind because we are curious about our own origins. There are two basic views concerning the origin of man. The papers of Charles Darwin and Alfred Wallace were read, jointly presenting their natural selection theory of evolution in 1858. A difference later developed between Darwin and Wallace on the question of man's origin, especially the origin of the human brain. Wallace believed that the human brain was the result of intelligent design. Darwin disagreed. Darwin extended his natural selection theory to the origin of human beings. His view that man evolved by natural means from some nonhuman creature was presented in his book, *The Descent of Man*,[19] published in 1871. It was fraught with philosophical and religious implications. Darwin's view, which rocked the intellectual world in the 19th Century, has become widely accepted in the 20th.

Does the fossil record provide any evidence for either the Darwinian or the intelligent design view of man? It is interesting to note that in his book, *The Descent of Man*, Darwin did not cite a single reference to fossils in support of his belief in human evolution. Clearly his original idea of human evolution did not grow out of a study of human fossil evidence, but out of a previously held opinion about the origin of man. The same is true of many researchers today. Since Darwin's time, Darwinists have been searching for fossil remains to establish their belief that man evolved. Over many decades, fossil after fossil has been heralded as "vital to the unraveling of the evolution of man." It has thus been easy to assume that human evolution has been confirmed just by the confidence expressed by many biologists in the essential correctness of Darwinism.

Before looking at the fossils recently proposed to be human ancestors, it is well to keep a few points in mind. First, comparatively few hominid fossil remains have ever been found. The April edition of *Science 84* makes the following observation:

> In all the world there are only a few dozen such specimens, a modest showing made all the more valuable by the decades of searching and digging by hundreds of specialists. Dinosaur researchers may examine thousands of specimens. Extinct elephants can be found by the hundreds at a single site. But when it comes to the remains of our own forebears, the numbers dwindle dramatically. Hominids, it seems, were never very numerous in ancient times. Those whose bodies were preserved, at least as bones, are rarer still.[20]

Another point to keep in mind is that the fossil remains of hominids are not only comparatively few in number, they are usually very fragmentary. A single bone or part of a bone, or for that matter, bones found in different locations are often used to reconstruct whole organisms. This means that interpretations of hominid fossil remains, because of their fragmentary condition, are very difficult and quite tentative. In part, this explains why paleontologists frequently cannot agree on any one scheme of human evolution, and why there is often argument over the classification of a given specimen. Let's turn now to an examination of the fossil record to see what light it sheds on the question of human origins.

Taxonomists place man in an order of mammals called *Primates*. This order is composed of animals that usually have their eyes directed forward and possess nails instead of claws. The standard fossil record of the primate order is shown in Figure 4-12. The fossil record of primates contains a variety of distinct types which appear abruptly, remain essentially unchanged, and in some cases abruptly disappear from the record!

Since the time of Darwin, there have been numerous candidates proposed as the "missing link." One by one, each of these fossil hominids has enjoyed a brief career, and then has been quietly set aside, for a variety of reasons. A recent example is *Ramapithecus* (Ram-a-PITH-i-kus), which until 1979, was placed by many anthropologists in the lineage leading to man. However, in 1979, a crushed but complete skull of a Ramapithecine was discovered in the foothills of the Himalayan mountains. This single reconstructed skull was able to indicate that *Ramapithecus* is more similar to the modern orangutan than to man, and consequently has been dropped by most anthropologists from the assumed ancestry of mankind.

What about the specific origin of man according to contemporary Darwinian thinking? Figure 4-13 shows two evolutionary sequences each favored by many Darwinists. The consecutive rectangles represent numerous hypothetical transitional forms for which there is no evidence in the fossil record. The numbers next to each known fossil give the estimated average cranial (brain case) volume in cubic centimeters. When reading this chart you may wish to translate the scientific names into plain English. Here is some help. For example, the root, *pithecus*, simply means ape. *Australopithecus robustus* (Awe-stray-lo-PITH-uh-kuss ro-BUS-tuss) translates simply as "the large (robustus) ape (pithecus) from the south (Australo)."

Prior to the fossils shown on the chart, there is a time period of many millions of years for which there are numerous presumed transitional forms and much speculation, but no fossil evidence. On the upper left of the chart we see a group of fossils with the genus name *Australopithecus* (called "australopithecines" as a group).

The first of the australopithecines, or "southern apes," was discovered in 1924 in South Africa by Raymond Dart. This group contains great variety among its members, and at different times different ones have been proposed as an ancestor of man. The best known form in the fossil record is *Australopithecus africanus* (see Figure 4-14 for a comparison with modern man). It has an average cranial capacity of about 500 cc (compared to an average of about 1400 cc for modern man), stood about 3½ to 4 feet tall, and weighed an average of 90 to 100 lbs.

Some Darwinists now consider only one of the australopithecine group a candidate for man's ancestor, "Lucy" (*Australopithecus afarensis*).

There is disagreement among Darwinists as to whether man's ancestry should be traced back to the australopithecines (specifically Lucy), or whether it should bypass this group and branch off from some hypothetical ancestor before the australopithecines. Donald Johanson and his co-workers who discovered Lucy (named after a song by the Beatles that was playing in camp the night of her discovery) hold to the first possibility. Richard Leakey favors the second.

Next, let us meet *Homo habilis* (HA-buh-lis), sometimes called the "tool maker." As you can see on the chart, the route to *Homo habilis* is controversial, and there is no agreement as to who, or what, is its ancestor. *Homo habilis* is the earliest species that has been placed in the genus *Homo*. The fossil was classified as *Homo* because of certain morphological characteristics, and this interpretation was encouraged by the conditions under which it was found. It was excavated in 1964 from a site littered with stone tools and animal fossils such as pigs, horses, catfish and tortoises. Was *Homo habilis* really the earliest human being, or was it only a primate like the australopithecines? Many Darwinists think it should be classified with the australopithecines. Some design proponents point out that its brain case, while larger than most australopithecines, is really too small for it to be classified as human (650 cc compared to about 1400 cc for modern man). They interpret it instead as an extinct primate. In fact, recent examination of its finger bones has led some scientists to conclude that the *Homo habilis* hand was similar in overall structure to chimpanzees and female gorillas.

Much is made of the presence of primitive tools near some *Homo habilis* finds. Living apes exhibit the opportunistic use of available materials as tools. For example, chimps will use sticks to probe a termite mound for food. There is nothing in the intelligent design view inconsistent with this use of naturally occurring materials, or even chipping stones, breaking sticks, etc., for simple purposes. At any rate, the presence of tools in the vicinity of the *Homo habilis* fossil might be explained by the discovery of human fossils that were found in

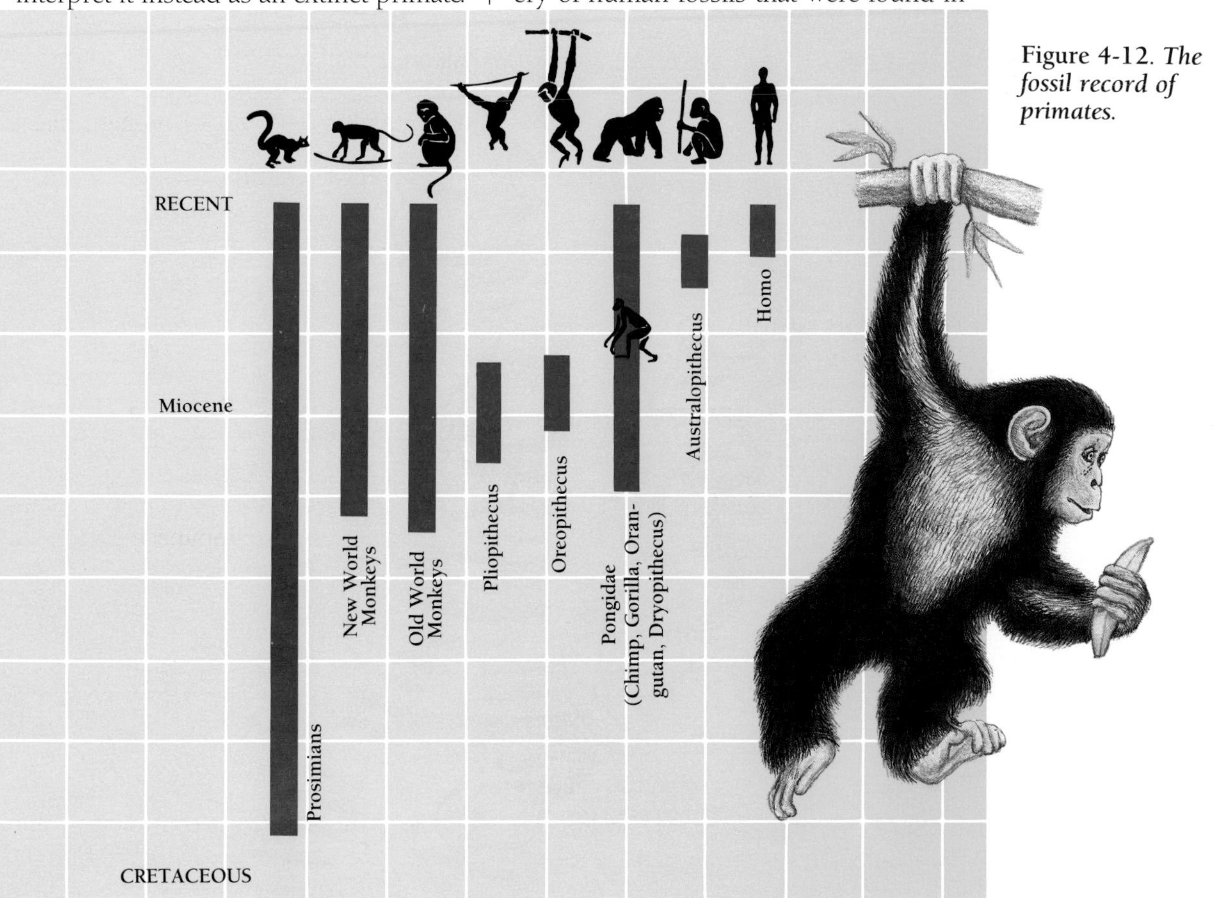

Figure 4-12. *The fossil record of primates.*

the same area by a later excavation.

Homo erectus had a larger brain (950 cc—see Figure 4-13) than *Homo habilis*, and walked with an upright posture, like man. Scientists date its appearance at about 1.8 million years ago. Along with *Australopithecus boisei* and *Australopithecus robustus*, it lived on the African savannah. It had significant anatomical differences from modern man that have prevented its classification as *Homo sapiens*. It also left no evidence that it buried its dead, no signs of art, or other recognizably human culture. Although Darwinists are divided on what lineage led to *Homo erectus*, they are largely united in their interpretation that it was an ancestor to man.

Less attention has been given to the first appearance of morphologically modern humans in Africa and the middle East, because of their recency, but these may be quite important discoveries. These earliest anatomically modern types were found at burial sites in caves at Qafzeh in the far south of Israel, and Es Skhul, for example. These and other fos-

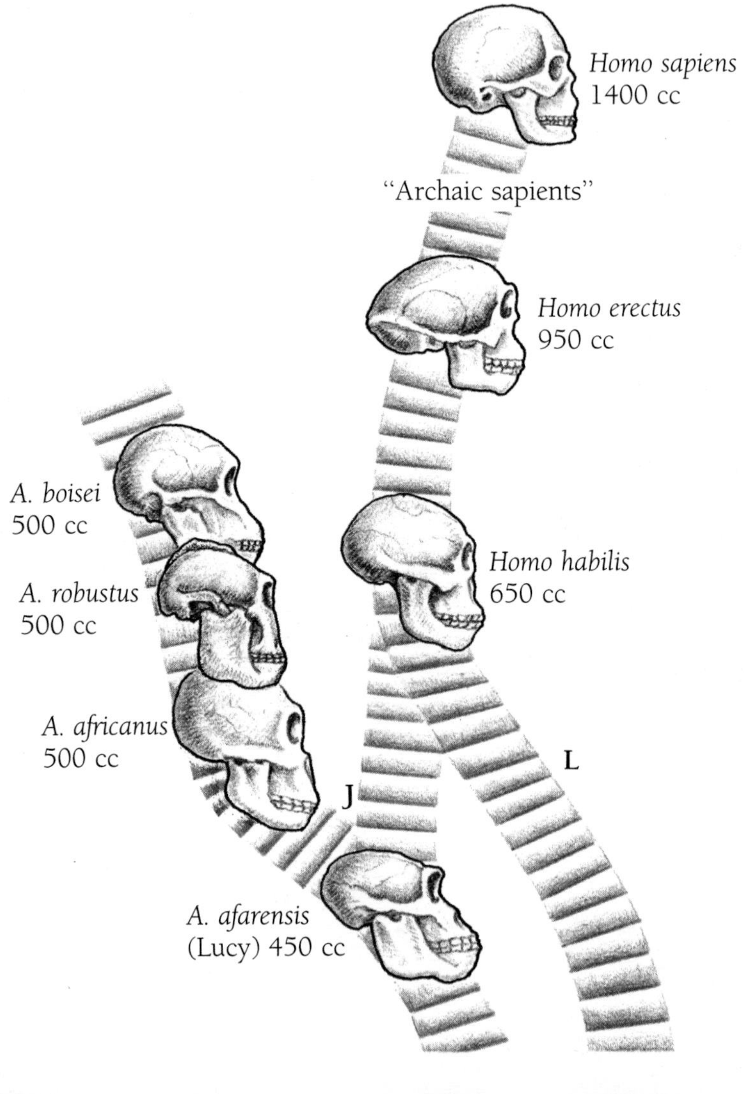

Figure 4-13. ***Presumed fossil sequence leading to modern man. Rectangles indicate gradualistic view; an unknown number of transitional species are not yet found in the fossil record. J=pathway according to Johanson. L=pathway according to Leakey.***

sils of modern humans dated as far back as 70,000 to 110,000 years ago. They had a clearly identifiable culture, with burial goods found in the burial sites. As early as 75,000 years ago, the "Hawieson Poort" industry in Africa featured tools that included handles securing tiny interchangeable blades. As this shows, their culture underwent continuous change through time. As David Wilcox put it, "In less than half the tenure of the Neanderthals, Cro-Magnon man was walking on the moon!"[21]

The best information we can seek about man's ancestors is that which tells us, not what they looked like, but what they did and how they behaved. Of course, such information is often very fragmentary if found at all, but it is the most important data for which to search. Without culture, relating different fossil types to one another is risky business (with high stakes, we might add, in our own culture). Even professionals are beguiled into accepting unwarranted conclusions. In the article, "The Descent of Hominoids and Hominids," David Pilbeam acknowledged that he was forced to lay aside his ideas of man's lineage:

> Why was the hominoid fossil record misinterpreted, at least by dimmer paleontologists such as me? There are a number of reasons. First, far too much attention was being paid to the fossil record as a source of information about evolutionary branching sequences. It is now clear that the molecular record can tell more about hominoid branching patterns than the fossil record does.[22]

While Pilbeam had in mind an application of molecular comparisons to primates, a fascinating and extensive molecular study was done on humans. Known as the "Mitochondrial Eve" thesis,[23] its publication in the mid-eighties brought a great deal of publicity. (The

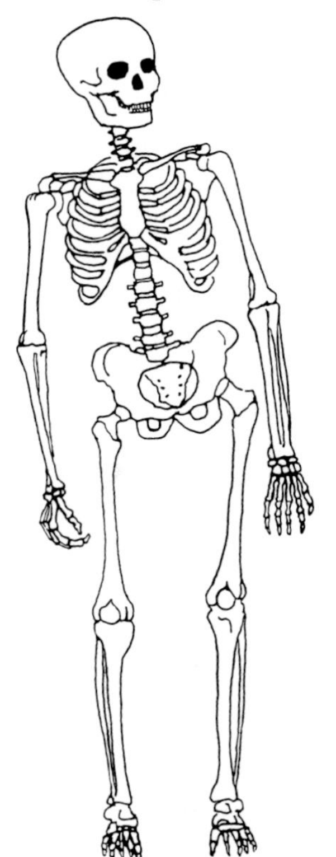

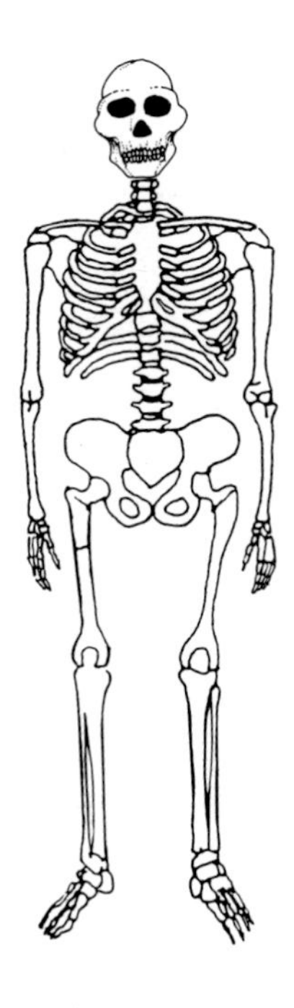

Figure 4-14. ***Reconstructed skeleton of <u>Australopithecus africanus</u>. A skeleton of modern man is included for comparison.***

name "Eve" was chosen as symbolic of humanity's first mother; there was no intent to identify this female with the biblical Eve.) The theory resulted from a study of widely distributed human mitochondrial DNA. It indicated that all races of humanity owe their origin to a single female ancestor in Africa, who dates to as recently as 200,000 years ago. (Various other dates have been suggested, some as low as 100,000, as criticisms and parallel studies have employed different methods of calibration.) The African location has been questioned as well.

The Mitochondrial Eve thesis is still very controversial, and involves a more general dispute between scientists of different disciplines (physical anthropologists versus molecular biologists).

The basis for the study is that mutations in the mitochondrial DNA provide a good means of tracking the relationships among peoples. The reason they do is that such mutations are not corrected as they are in nuclear DNA. Therefore, comparable segments of the DNA from two populations can be compared, allowing investigators to detect any ancestor/descendent relationship between them. Such sequence comparisons can help in determining the relationships of the various races of humanity going back in time. They also give us, if couched in some necessary assumptions, a means of measuring this "molecular root" of humanity. As we have said, it measures back to 100,000 to 200,000 years. Though startling to scientists, a "molecular root" no longer than this seems to fit very nicely with the fossils of modern humans dating approximately 110,000 years ago.

If the theory turns out to be confirmed in some reasonable approximation of its current form, it would have three major implications in man's quest for his ancestry: 1. It would mean that humanity, as represented by its contemporary peoples, is dramatically younger than traditionally conceived by most scientists. 2. It would eliminate Neanderthal as a candidate for ancestry to European peoples. 3. It would eliminate the vast majority of *Homo erectus* populations across Europe and Asia as ancestral to man, leaving open only the possibility that some tiny *Homo erectus* population in Africa gave rise to modern *Homo sapiens*, if the two are related at all.

As in the case of mammal-like reptiles, hominids present us with several eligible candidates for evolutionary transition. The mere existence of these hominid fossils carries a certain power of suggestion. But Yale University Professor of Biology Keith Thomson has given a very relevant caution:

> Although "finding ancestors" is the traditional paleontologists' "proof," such "historical events" cannot be tested by assembling nice series of fossils without discontinuities, because the evolutionary hypothesis is superficially so powerful that any reasonably graded series of forms can be thought to have legitimacy. In fact, there is circularity in the approach that first assumes some sort of evolutionary relatedness and then assembles a pattern of relations from which to argue that relatedness must be true. This interplay of data and interpretation is the Achilles' heel of the second meaning [descent through common ancestry] of evolution.[24]

Who, or what were man's ancestors? The fossils surely don't give us any conclusive answer. Darwinists are convinced that *Homo erectus* was nearly human, and directly ancestral to man. Design adherents, however, regard *Homo erectus*, as well as the other hominids discussed in this section, as little more than apes, and point instead to the abrupt appearance of the culture and patterns of behavior which distinguish man from the apes.

Suggested Reading/Resources

Darwinism: The Refutation of a Myth, by Soren Loutrup. New York: Methuen, 1987. This refutation of Darwinism includes a worthwhile discussion of the fossil record.

References

1. T. Dobzhansky. *American Scientist*, Dec. 1957, p. 388.
2. C. Darwin, 1968. *The Origin of Species*. Penguin Classics edition, London: Penguin Books, p 292.
3. R.A. Raff and E.C. Raff, eds. 1987. *Development as an Evolutionary Process*. New York: A.R. Liss, p. 84.
4. S.J. Gould, 1989. *Wonderful Life*. New York: W.W. Norton. p. 64.
5. D. Erwin, J.W. Valentine, J.J. Sepkoski, Jr., 1987. *Evolution* 41,1179.
6. D. Raup, 1979. "Conflicts Between Darwin and Paleontology," *Field Museum of Natural History Bulletin*, 30 (1), 25.
7. S.J. Gould, 1977. *Natural History* 86, 14.
8. Ibid.
9. S.M. Stanley, 1979. *Macroevolution: Pattern and Process*. San Francisco: W.H. Freeman, p. 82.
10. H.C. Bold, 1967. *Morphology of Plants*. New York: Harper and Row, p. 515.
11. J.A. Hopson, 1987. *American Biology Teacher* 49,16-26.
12. Ibid., p. 19.
13. D.J. Futuyma, 1983. *Science on Trial*. New York: Pantheon. p. 85.
14. P.D. Gingerich, B.H. Smith, E.L. Simons, 1990. *Science* 249,154-157.
15. P. Wellnhofer. *Scientific American*, May 1990, pp. 70-71.
16. D.L. Wilcox, 1993. *The Creation: Spoken in Eternity; Unfolded in Time*. In press. p. 6/23.
17. B.J. Stahl, 1985. *Vertebrate History: Problems in Evolution*. New York: Dover. pp. viii, 369.
18. C. Patterson, in Luther Sunderland, 1981. *Darwin's Enigma: The Fossil Record*, ED 228 056, p. 21.
19. C. Darwin, 1871. *The Descent of Man*. New York: Modern Library.
20. B. Rensberger. *Science 84*, April 1984, p. 29.
21. D.L. Wilcox, 1993. *The Creation: Spoken in Eternity; Unfolded in Time*. In press. p. 7/12.
22. D. Pilbeam. *Scientific American*, March 1984, p. 87.
23. A.C. Cann, et al., 1985. *Biological Journal of the Linnean Society*. 26,375-400.
24. K.S. Thomson. *American Scientist*, Sept./Oct. 1982, pp. 529-530.

EXCURSION CHAPTER 5

Homology

Introduction

One of man's most ancient observations of the biological world is the repetition of various structures. Frogs, rhinoceros, iguanas, and camels are quadrupedal, having four limbs. Whales, fish, sea lions, and manatees have flippers. Squids, lemurs, snakes, and giraffes have eyes; and stegosaurs, porpoises, and fish have dorsal fins. Explanations of these similarities among diverse organisms have risen and fallen over time. Today, the Darwinist assumes that similar structures stem from the common ancestry of all living things. The concept of **homology** applies to biochemistry, physical structure, and physiology, as well as to behavior and often, function.

Even before the rise of Darwinism in the 19th century, the concept of structural homology was used to describe the relationships of body parts in different organisms. For example, the hand of a man and the hand of a monkey were said to be homologous. This is particularly easy to understand since the two parts both look alike and perform somewhat the same function.

The Darwinian definition of homology, however, does not rest upon the two parts being identical, or even similar for that matter, in appearance or function. Darwinists have defined homology as correspondence of structure derived from a common primitive origin, a definition that assumes macroevolution to be true. To take one example, the human hand and the dog's forepaw contain the very same types of bones, and the bones of both have been given the same names (see Figure 5-1). Although the bone pattern is similar, the individual bones are quite different in shape and function. The dog cannot grasp objects with its digits or oppose its thumb to the rest of its toes, either individually or collectively. Human hands, of course, are not made for walking or standing for any length of time. Their similarity to dogs' paws does not result from function. The only alternative, according to the Darwinists, is that their similarity results from a common genetic inheritance from a common ancestor that possessed this basic arrangement of bones. To the Darwinists, human hands and dogs' paws are homologous.

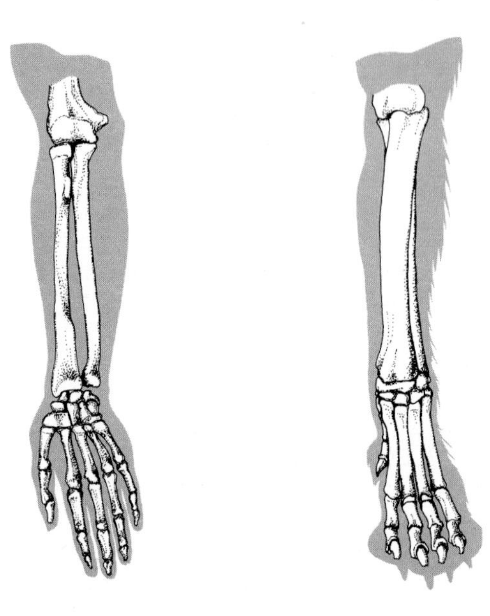

Figure 5-1. ***The front appendages of a dog and a man. Notice the bones in each correspond to the bones in the other.***

What was the common ancestor? Darwinists assume it was some type of *insectivore* (the order of mammals that includes moles and shrews). This insectivore in turn is assumed to have evolved from an ancestral reptile which in turn received its forelimbs from a semiaquatic fish. The fish presumably used its bony fins to drag itself from mudhole to mudhole during seasonal droughts. A fin might not seem the best pattern to use for a horse's leg or a dog's paw, but it was the only raw material available. According to this view, both a man's hand and a horse's hoof are homologous to the fin of a fish. There were no new ideal parts available, so Darwinian mechanisms fashioned and remolded existing parts. If one uses a little imagination, an equivalent of each fin bone should be identifiable in the human hand and the horse's hoof.

One goal of Darwinists is to trace the evolutionary history of fossil organisms. While they have recognized many reasons that gaps should exist in the fossil record, they have long expected that fossils would ultimately be found to document the theory of "descent with modification." Yet continuous evolving series of fossils are never found. Necessary common ancestors are missing in the fossil record, creating large gaps between the major groups of organisms. Therefore, the only means left to determine relationship is the concept of homology. Scientists have hoped that as the fossil remains of two organisms are examined, an adequate judgment can be made concerning how many homologous structures they have in common. Other scientists, however, including Darwinists, are very skeptical about the value of fossil remains in determining phylogenies.[1]

Because it applies to such a wide variety of life forms, the concept of homology has become a central pivot upon which much evolutionary argument turns. Darwinists reason that the more homologous structures two organisms share, the closer is their evolutionary relationship. If two structures are truly homologous in the Darwinian sense, they should have the same embryonic origin, and their development should be controlled by at least some of the same genes.

But are we always able to determine which structures are homologous? There may be very little similarity in appearance between structures, as is the case with a bat's wing and a horse's leg. Darwinists consider these structures homologous, even though there is no obvious similarity in pattern.

Homology or Analogy: A Problem of Interpretation

The Meaning of Analogy

Structures considered homologous may look different and function differently, but still be considered homologous to satisfy Darwinian theory. Very similar structures may not be considered homologous, although they perform the same function. Structures having a similar appearance and function in unrelated groups are said to be *analogous*. In other words, Darwinian mechanisms have arrived at similar solutions to the same physiological problem using very different raw material. Let us look at an example of analogy.

Figure 5-2 shows the skull of a dog next to that of a Tasmanian "wolf" and a North American wolf. Tasmania is a large island adjacent to Australia that, like Australia, contains a large variety of marsupials. The Tasmanian "wolf" is a marsupial which in general appearance and behavior is very similar to the *placental* wolves found in other parts of the world. Even the behavior of this now extinct animal was similar. The Tasmanian wolves ate the settlers' livestock, and as a result were hunted until they became extinct. But although they behaved like placental wolves, a study of their anatomy suggests that Tasmanian wolves were actually more similar to kangaroos. Darwinists interpret the anatomical findings to indicate that the two types of wolves are only remotely related, and that each had a separate evolutionary history since the time when the Australian continent was separated from the continent of Antarctica. Yet the skulls of the two wolves are extremely similar, as you see. How did this come about?

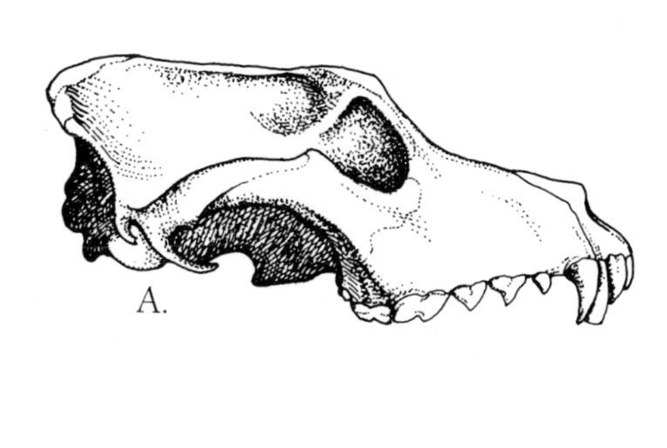

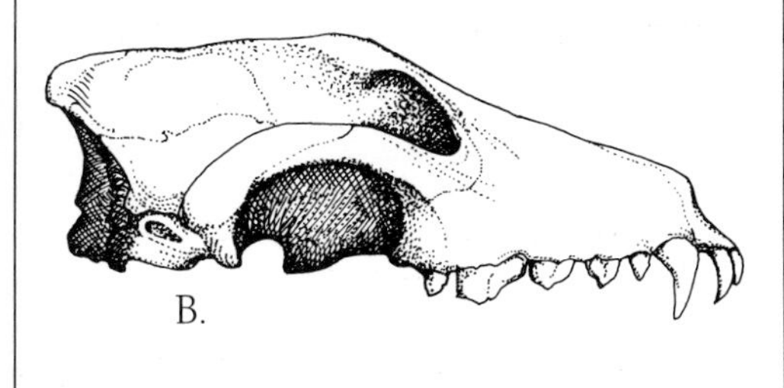

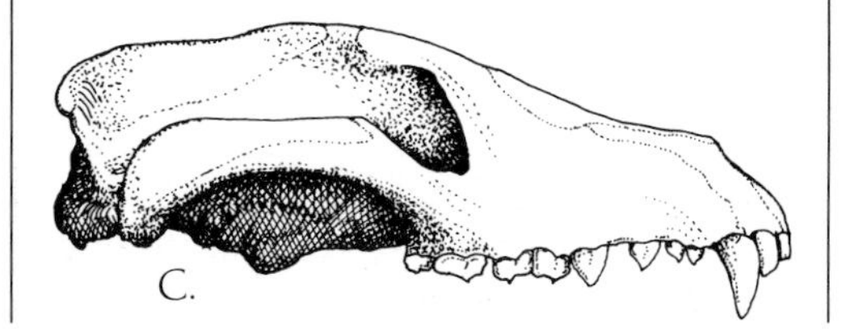

Figure 5-2. *The skulls of a dog (A), a North American wolf (B), and a Tasmanian wolf (C). Notice that the skull of the North American wolf is somewhat similar to the dog's, which is said to be related to it, but nearly identical to the Tasmanian wolf, which is allegedly only distantly related to it.*

According to Darwinists, both groups evolved into wolflike forms, an occurrence known as **convergent evolution**. This is a form of coincidence; it means that two lines of descent took different evolutionary paths that finally converged, having independently developed similar features adapted to meet the same environmental demands. Apparently the selective regime that produced the North American wolf was established by niches closely approximated in Australia, so that the two approached this ideal ever more closely with the passage of time, increasingly coming to resemble one another until they became superficially almost identical. As time passed, Darwinian evolution, through chance experimentation, had independently developed the same general forms in two different areas of the world. Examination of the two wolves' skulls would lead us to wonder just which features were homologous and which were analogous.

If two organisms are judged to be related through a "recent" common ancestor, their similarities are said to result from **parallel evolution**. But the wolves reflect the concept of convergent evolution (see Figure 5-3). Another example of this concept is the vertebrate eye

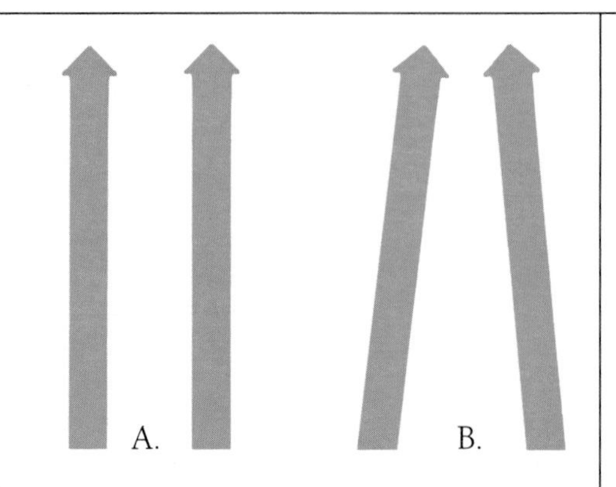

Figure 5-3. ***Convergent and parallel evolution. When similar organisms evolve similar structures independently, parallel evolution (A) is said to have occurred. When this occurs between distant taxa, it is called convergent evolution (B).***

and the eye of the squid. (This same idea, on a cosmic scale, is portrayed in many of the space movies which show extraterrestrial intelligence in man-like forms.) In some cases very general similarities are taken as evidence of evolution (homologies), whereas in this case impressive similarities are not. How can we know when similarities indicate kinship? In order to decide, other considerations must be used, such as the geographical history of the Australian continent and the significance of other anatomical details. Darwinists and proponents of intelligent design alike agree that such clues should be sought, recognizing that finding and evoluating this information is extremely difficult.

The Panda Connection

Behind that favorite character of the zoo animals, the giant panda, there is a fascinating story of scientific investigation. The giant panda is native to the bamboo mountain forests of southwest China, as is the lesser, or red panda (see Figure 5-4).

For over a century, those who have studied the two pandas have been unable to agree on whether they are members of the bear family or of the raccoon family. Since the first serious attempt to classify them in 1869, more than 40 major scientific studies have been published on the subject. The astonishing thing is that they have been split almost down the middle on the bear/raccoon question, half concluding they are bears, half concluding they are raccoons. One scientist called the phenomenon a "taxonomic game of ping pong."

Then in 1964, Dwight Davis, Curator of Vertebrate Anatomy at the Field Museum of Natural History in Chicago, published what soon became widely accepted as the definitive discussion on the matter, which finally solved the mystery of the pandas.[2] Davis had concluded, and most agreed, that the giant panda was not a raccoon but a bear, and that the red panda was not a bear but a raccoon! More recently, studies of accumulating biochemical data have extended the list of similarities between the giant panda and other bears.

But why, then, were biologists and others convinced for so long that the two pandas were close relatives, both in the same family? One reason is mere geography. If the red panda is a raccoon and the giant panda a bear, then the red panda has many family relatives but none outside the Western hemisphere. It is easy to understand why this may have helped to mislead biologists trying to classify the pandas.

Besides lumbering around the same neighborhood, the giant and red pandas share an impressive number of behavioral and physical traits in common. These traits are unique to the pandas; they did not receive them by descent from their respective ancestral families. For instance, the muzzle or snout of each is similar in shape, as are their maxillae (upper jaws), shown from below in Figure 5-5.

Notice how short the muzzles of both are in comparison to the polar bear, and how sharply the jaw bones widen toward the back of the head. In coordination with these traits, both pandas have massive premolar teeth and enlarged chewing muscles. These similarities are obvious to even

Figure 5-4. *Giant panda and lesser, or red panda.*

casual observers, but other notable ones are not so apparent.

The two pandas stand apart from the bears with several characteristics of the stomach and alimentary tract more subtle than those of the jaw. The same is true of comparisons of their livers. We might ordinarily expect that these characteristics arise from the bamboo diet the two pandas share, but at least one contemporary scientist claims they do not fit this idea. Genetically, the giant panda has much in common with the other bears, yet it has 42 chromosomes, far closer to the red panda count of 36 than to the 74 chromosomes of most bears.

The giant panda, which needs no additional reasons for its fame, has become known for the "thumb" which gives it a dexterity not found among other bears. This structure operates like an opposable thumb, although it is not a true thumb and is only partially opposable (see Figure 5-6). Actually, it is an enlarged bone of the panda's wrist, known as the radial sesamoid (SES-a-moyd). The cluster of bones in the wrist must function together smoothly, many of them surface to surface, through all of the paw's manipulations—opening, closing, swiveling, swatting, etc. Thus the details of the surface contours are important to the paw's functions. The panda is known for its

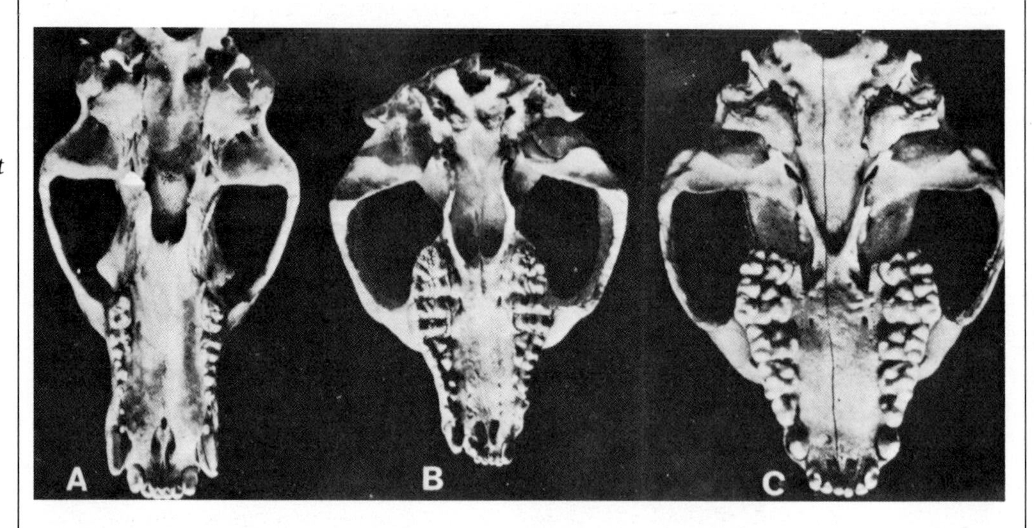

Figure 5-5. ***The upper jaws of (A) polar bear, (B) giant panda, (C) red panda.***

ability to handle and strip bamboo, an activity that consumes a major part of its day and for which the enlarged radial sesamoid suits it unusually well. Even here, the two pandas share similarities quite apart from raccoons and bears. One similarity is the enlargement of the radial sesamoid, seen even in the red panda, although to a lesser degree. The red panda even uses it to handle its bamboo dinner in a way similar to the giant panda.[3] A second similarity is the special way the tendon from the abductor muscle fits into the radial sesamoid, and a third is the detail of a working surface among the cluster of wrist bones. Such detailed similarities would make great material to support a common ancestry argument!

In addition, the red and giant pandas have many similar behavioral characteristics. For example, unlike most other bears, neither one hibernates. In some ways, the behaviors of the pandas are parallel, and in other ways, behavior patterns of the giant panda differ from those of other bears. Curiously, the giant panda does not growl, bark or roar like the bears. Instead, it bleats, much like a sheep. As you can see, there is a rather impressive list of characteristics shared by the two pandas. Even though biologists have learned about these progressively over the last century, it is easy to see why early in the scientific study of pandas, these characteristics were thought to be homologies, and thus why biologists classified the pandas together in one family.

But suppose we accept the view that currently predominates among biologists, that these unique features shared by the pandas are not homologous, but rather

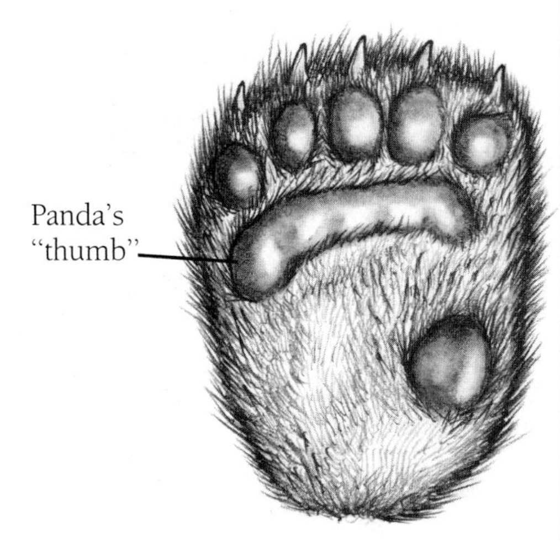

Figure 5-6. ***The panda's "thumb." Actually an enlargement of a bone in the wrist, located where the panda can use it with his claws to grip objects.***

analogous—examples of convergent evolution. Is this an exception, or the rule? When do we trust the features at hand to inform us truly about evolutionary relationships? This question is all the more striking when we consider the Panda example alongside a well accepted example of homology.

Linking Classes Together with Homologies

We saw in Chapter 4 that the first mammals are thought to have come from reptiles known as Therapsids. The general body plan of short-legged quadrupeds is shared by many reptiles, amphibians, and mammals. Yet many of the evidences for the major changes necessary to transform a reptile into a mammal relate to the soft-bodied parts. Soft-bodied parts are not usually fossilized, and these evidences are also missing from the fossil record.

What homologies, then, tell Darwinian scientists that mammals evolved from Therapsids? Notably, the skulls and mandibles (lower jaws) of the Therapsids are said to have bones homologous to those of the first mammals. The upper and lower jaws of reptiles articulate (fit together) with two bones (one each located at the back of each jaw) not found in mammals. According to Darwinian theory, these two bones have become relocated in the middle ear of the mammals through evolutionary descent (see Figure 5-7). Yet there is no fossil record of such an amazing process. Consider that to make this change, one of these bones had to cross the hinge from the lower jaw into the middle ear region of the skull, where Darwinian mechanisms reshaped and refined them into highly specialized,

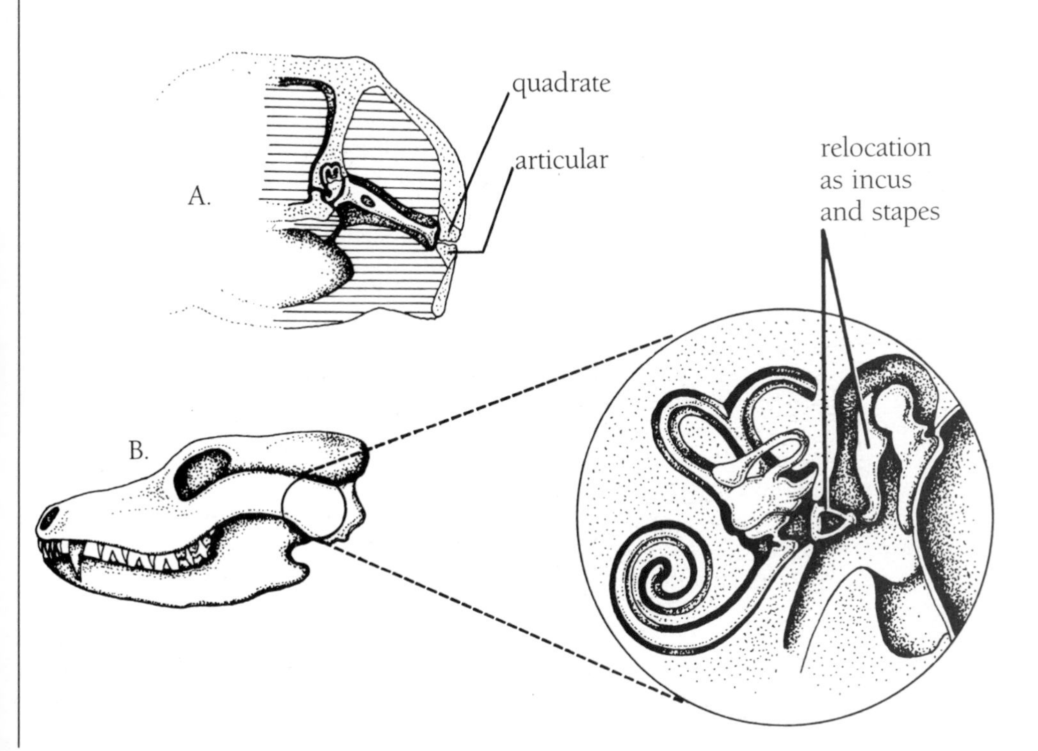

Figure 5-7. *The location of bones, considered homologous in reptiles and mammals. The skull pictured in (A) is sectioned through the back to show the small articular and the quadrate bones located where the upper and lower jaws meet. (B) shows their relocation according to Darwinian theory, to the middle ear of mammals. (After Romer)*

delicate instruments of sound transmission. Such an occurrence would be extraordinary enough by itself, but some Darwinists propose that this happened more than once!

This does indeed balance the count of bones between the two, but can the "homology" of an identical bone count serve as reliable evidence for the macroevolution of mammals from reptiles? If it can, then how can we dismiss the far more detailed similarities between the two pandas and declare they do not indicate evolutionary relationship? How can we have it both ways? Can we be confident of evolutionary relationships based on remote similarities if large groups of striking similarities like those between the pandas do not indicate kinship?

Homologies and the Fossil Record

Darwinists assume that living organisms are related to one another through common descent if they possess homologous structures. The more homologous structures they have in common, the greater their degree of relationship, according to this line of thought. The same reasoning is applied to extinct organisms known to us only through the fossil record; here scientists make use of the concept of homology to bridge the gaps in the record. In this area, too, the number of homologies between organisms guides their judgments about the relatedness of fossil types.

But these judgments must be even more subjective than for living organisms because there is normally no way of determining embryological development or genetic similarities from a fossil. Rare exceptions include some fossil dinosaur embryos recently found in fossilized dinosaur eggs on Egg Mountain in northern Montana[4] and the mitochondrial (MY-doh-KOND-re-uhl) DNA of some trees dated to 20 million years ago.[5] Evolutionary paleontologists usually have only general external anatomy to work with, especially with mammal fossils. So they are left to establish relationships based upon a lot of speculation. Not only is the concept of homology ambiguous, but the data from the fossil record are insufficient in determining degrees of relationship among fossil organisms. Convergent and parallel evolution may be convenient explanations for some similarities, but they are not very persuasive when innumerable forms—the intermediate steps of the two converging lines of descent—are missing.

For the design proponent, there is another explanation of the origin of analogous features in unrelated groups. The design proponent assumes that the similarity of features can be accounted for on the basis of *design requirements*. In other words, the skulls of marsupial wolves and of placental wolves are similar because one particular skull design best suited the requirements of both organisms. The pandas share many characteristics in common for the same reason. We call this idea *teleology*: an organism is designed for certain functions or purposes. Evolutionists avoid this concept since, in the past, it was unquestionably overstated with pat answers about design, serving to squelch further inquiry.

The Concept of Homology: Limits of its Usefulness

Subjective Determination of Homologies Among Living Organisms

The determination of homologies among groups of living organisms is also a matter of subjective judgment. For an example, let us return to Australia. Besides wolves, we find many other marsupials that resemble placental mammals found in North America, such as flying squirrels, mice, moles, and woodchucks (see Figure 5-8). Yet the Australian marsupials do not reproduce in the same way that North American wolves, moles, and woodchucks do. The North American versions use a placenta that allows the female to nourish its young while the babies are still in the uterus. The Australian versions have an external pouch (*marsupium*) for holding their young, which are born in a very immature, almost embryonic state. (One South American animal also has such a pouch, the opossum.) So we are faced with questions such as: Is the Australian mole more closely related to its North American counterpart, which resembles

Figure 5-8. *A woodchuck and his marsupial counterpart, the wombat.*

it but lacks a pouch, or is it more closely related to the opossum which does not resemble it but which has a pouch? (A fascinating variation occurs in the position of the pouch of the Australian wombat. In contrast to the pouches of most marsupials, which open at the top, it opens at the bottom. This is a merciful relief to the young wombats since, when the mother is tunneling underground, her forepaws would certainly pack the pouch full of dirt were it otherwise!) Most marsupials also possess a small pair of bones—*marsupial bones*—associated with the pelvis, and supporting the pouch, yet they are never found in placental animals. Darwinists classify the pouch and marsupial bones as homologous in the various marsupials.

It is becoming evident, then, that the determination of relationship rests upon a subjective judgment. If only one set of similarities existed, they would undoubtedly be regarded as homologous by Darwinists. In this case, which set of similarities will be a reliable guide: similarities in reproduction, or in general appearance? Notice that the answer depends upon the similarities that one picks as being most significant; that is, upon one's presuppositions. Darwinists choose the similarities in reproductive structures as being the most important. These features are considered to be homologous, indicating close relationship, while the similarities in general appearance (such as the two versions of woodchucks) are considered to be analogous, the result of parallel evolution. In other words, marsupials are thought to have evolved in a line of descent parallel to, but independent of, placentals.

Darwinists reason that natural selection is more likely to produce similarities in general appearance of unrelated groups like marsupials and placentals than it is to produce specific structures such as pouches and marsupial bones in unrelated groups. Why? Because there is no obvious reason why natural selection would have produced these reproductive structures in unrelated animals (if the various marsupials are unrelated), but there is a reason why natural selection might have produced similar generalized body forms in certain unrelated animals. General appearance, such as the build and teeth in a woodchuck, are more advantageous to unrelated organisms adapted to the same life style and similar environment. In the case of a woodchuck and a wombat for instance, a heavy body build and gnawing dentition would be helpful to both (see Figure 5-9), and it is easy to say similar things about the different types of moles and wolves. But if all of the Australian marsupials with their varied life styles have descended from a common ancestor that had marsupial bones—bones which are irrelevant to the distinctive life styles they each lead—then those marsupial bones will probably be retained in all of them. This would suggest that they hold some genes in common, and that they are related to one another.

Homology Requires Assumptions

But consider how limited the usefulness of the concept is; it cannot replace subjective judgments, nor can it deliver us from the influences of our presuppositions. Notice that the truth of Darwinism must be assumed before such an assessment of relationship can be made. Marsupial pouches and bones are considered homologous because it is assumed that the organisms possessing them descended from a common ancestor. It cannot be denied that this interpretation is consistent with Darwinian theory; yet the consistency may be merely because we assume they are related, and nothing more. This is a classic example of circular

Figure 5-9. ***Like the woodchuck, the wombat is a powerful chewer, and possesses gnawing incisor teeth.***

reasoning. Stated differently, the harmony of Darwinian evolution and a homologous interpretation for these structures is certainly worth our notice. But in the light of a second set of similarities, such as those shared by placental and marsupial wolves, placental and marsupial moles, etc., the existence of these pouches and bones in Australian mammals merely poses a riddle. They are not actually evidence that macroevolution did occur; for it is an ancient and valid principle of logic that one cannot assume, for argument's sake, the truth of a proposition as a means of proving it.

The Intelligent Design Alternative

Is there any alternative explanation for the marsupial bones and pouches other than that they are homologous and therefore evidence for common ancestry? Yes, another theory is that marsupials were all designed with these reproductive structures. But, the question might be asked, Why were not the North American placentals given the same bones? Would an intelligent designer withhold these structures from placentals if they were superior to the placental system? At present we do not know; however, we all recognize that an engineer can choose any of several different engineering solutions to overcome a single design problem. An intelligent designer might reasonably be expected to use a variety (if a limited variety) of design approaches to produce a single engineering solution, also. Even if it is assumed that an intelligent designer did indeed have a good reason for every decision that was made, and for including every trait in each organism, it does not follow that such reasons will be obvious to us. These questions can, nevertheless, generate research in areas we might never investigate if we believe no answers exist, as we might if we are convinced that undirected or loosely directed chance produced both (marsupial and placental) structures.

One feature of *The Origin of Species*, persuasive in Darwin's day, was that it offered an alternative to design in nature. In it, natural selection was unveiled as a mechanism that would show why design is really only apparent design. In the thirteen decades since *The Origin of Species* was published, however, considerable evidence has come to light which severely questions Darwin's views. As a result, the plausibility of apparent design has suffered, while the rationale for reconsidering real design has grown stronger.

Surely the intelligent design explanation has unanswered questions of its own. But unanswered questions, which exist on both sides, are an essential part of healthy science; they define the areas of needed research. Questions often expose hidden errors that have impeded the progress of science. For example, the place of intelligent design in science has been troubling for more than a century. That is because on the whole, scientists from within Western culture failed to distinguish between intelligence, which can be recognized by uniform sensory experience, and the supernatural, which cannot. Today, we recognize that appeals to intelligent design may be considered in science, as illustrated by the current NASA search for extraterrestrial intelligence (SETI) (see Figure 5-10).

Figure 5-10. ***SETI. NASA's large radio-astronomy dish at Goldstone, California is one of many engaged in the search for extraterrestrial intelligence. (SETI), "rummaging through the dark crannies of space," looking for evidence of intelligent cause.***

Archaeology has pioneered the development of methods for distinguishing the effects of natural and intelligent causes. We should recognize, however, that if we go further, and conclude that the intelligence responsible for biological origins is outside the universe (supernatural) or within it, we do so without the help of science.

What Is Homology, Really?

Can we say with certainty what structures are homologous? One might think from the discussions found in most textbooks, that homologies are obvious. Often, however, that impression results from carefully chosen examples. When other examples—say, the ear ossicles—are examined, homology seems far less obvious. In some cases, fossil evidence of homology simply cannot be put forth. In others, the actual embryonic origin of supposedly homologous structures is different.[6] In still others, there is no evidence that the development of supposedly homologous structures is under the control of corresponding genes in the two organisms.[7] Then what, if anything, is homology? Some textbooks omit these important facts, or even contradict them, while still applying the principle to selected examples in support of Darwinian evolution. Many students, therefore, may accept Darwinian evolution as far better established than they otherwise would if given access to these facts. Despite the great importance of the concepts of homology and analogy in Darwinian thinking, it is not possible to apply them consistently, or even to be sure that, in certain situations, one or the other exists.

By now you can see why design proponents assert that the concept is too confusing to settle questions of structural classification. Structures are classified as analogous or homologous with no objective criteria for the choice. Homologous structures can only be identified with confidence if Darwinism is presupposed. But if one classifies organisms on the basis of assumed evolutionary relationships, how can that classification be used as evidence that evolutionary relationships are real? The reasoning is circular, the confidence misplaced. That means that neither the concept of homologous structures nor any concept derived from it can be used as evidence for Darwinism.

What if, indeed, the kinds of similarities that are basic to the concept of homology originated apart from Darwinian mechanisms? If they admit to two or more interpretations, can they be very compelling as evidence at all?

Among organisms classified as distant relatives, homology is an uncertain guide to the degree of their relatedness, and is of no value at all in proving that they are related. Design proponents have a realistic and more cautious approach to the use of homologies. They regard organisms which show great structural differences, such as starfish and chimpanzees, as having no common ancestry. The concept of homology as an indicator of evolutionary relationship cannot be applied to these organisms at all, because it would have no meaning if they were in fact unrelated. On the other hand, for organisms that are genetically related, such as the Hawaiian fruit flies, design proponents believe the concept of homology can be useful in determining relationships. They also point out that, for organisms showing greater differences, Darwinists apply the concept of homology very selectively and subjectively, when it can be applied at all.

The cases typically given in introductory textbooks are easy ones for Darwinian interpretation of homology. But it is the hard cases that test the power of an argument. We will show this with one of the easiest Darwinian cases and then go on to a hard one.

The Coccyx: An Easy Case for Evolution

Most vertebrates have tails, but a few do not. The guinea pig, gorilla, chimpanzee, and human being are tailless as adults. Using the concept of homology, Darwinists look for a skeletal structure in the tailless forms that is homologous to the tail found in other vertebrates. If such a structure is found, Darwinists would interpret it as evidence that these tailless forms descended from ancestors having functional tails. Does such a structure exist in human beings?

The answer given is, yes. It is the *coccyx* (KOCK-six), a short tail-like collection of three to four vertebrae attached to the end of the spine, but normally not noticeable from the exterior (see Figure 5-11). The coccyx was once classified as a **vestigial structure**, one that had a function in our presumed ancestors but which over time lost that function and has become reduced to a vestige in us. Otherwise, why would it be found in humans? ask Darwinists. Is the coccyx in man homologous with the tail of a dog? Your decision depends upon your prior belief about macroevolution. If you already believe that men and dogs came from a common ancestor, you will view the coccyx as evidence for that evolutionary relationship. However, there is also evidence that the coccyx has a different function in man. It serves as a point of attachment for several important muscles of the pelvic floor. While anatomical and medical textbooks describe this and other functions of the coccyx rather matter-of-factly,[8] many high school biology texts prefer to continue calling it a "vestigial structure." The same is true of the human appendix, which is now known to play a functional role in the immune system,[9] and several other "vestigial structures."

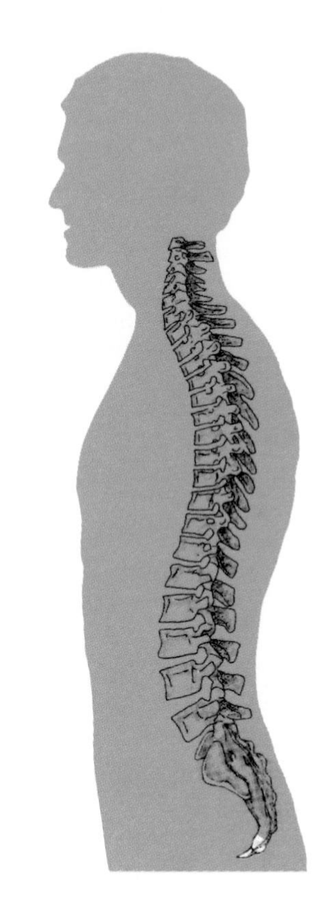

Figure 5-11. *The coccyx. The coccyx is composed of the 3-4 final vertebrae at the base of the spinal column, thought by Darwinists to be a remnant of a tail from man's evolutionary ancestry.*

Recapitulation: An Evolutionary Application of Homology

In 1866, Ernst Haeckel stated his now-famous *Biogenetic Law:* "Ontogeny [the embryological development of an individual] is a brief and rapid recapitulation [review] of the phylogeny [evolutionary history of the organism]."[10] This concept does not receive the attention from Darwinists today that it once did, although it hasn't been entirely dismissed.

As an example, consider how the Biogenetic Law has been applied to the human species. Human beings begin their embryological development as a single cell, called a *zygote*. The zygote, we are told, is comparable to our earliest supposed ancestors, the single-cell protists. The next stage of development, the *morula*, is a cluster of cells that is similar to colonial protists (colonies of individual but not entirely independent cells), especially the flagellated plant-like forms. (It is interesting to note that Haeckel coined most of these terms, which were therefore new to his opponents in debate. This was of considerable help to him in winning arguments.) Soon a cavity develops in the center of the morula and a hollow ball of cells results, known as the *blastocyst* (see Figure 5-12). This is supposed to be reminiscent of one of the larger ball-like, plant-like colonial flagellated protists such as *Volvox*. At this point human embryology is not a good example of Haeckel's law, but in some other animals the next stage is a *gastrula*, a double-walled structure that is a little like a cnidarian (ni-DARE-e-un) (a colonial marine animal similar to Hydra) in its basic architecture. Returning to human embryology, a stage is eventually reached in which there are structures resembling the aortic arches of fishes, and much later there is a fur-like coat, *lanugo* (la-NOO-go) which temporarily covers the fetus but disappears before birth. So is this supposed "retracing of evolutionary history" evidence for Darwinian evolution? Some would have us believe so. A number of current textbooks give partial presentations of this story, very often stating that gill slits appear in the human embryo. But this has been shown false by advances in embryology, and is uniformly rejected by scientific literature.[11] Yale University Biology Chairman, Keith Thomson, said, "The biogenetic law as a proof of evolution is valueless,"[12] and others have even termed it "a mirage."[13] Moreover, precise photographs of the various stages of the embryo's development show them to be at odds with the idea of recapitulation. The question now is, what relationship does homology have to this idea of recapitulation? Let's examine some specific cases of recapitulation to show how the concept of homology has produced faulty conclusions.

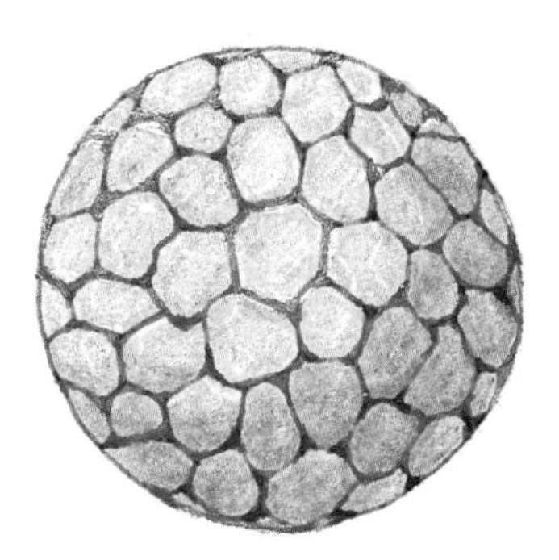

Figure 5-12. ***The blastocyst. One of the embryo's very early stages. Many Darwinists have thought the outward similarities between the blastocyst and the volvox (a round, hollow, plant form) represent a stage of recapitulation of the embryo's evolutionary history.***

The Kidneys: An Easy Case for Recapitulation

The human and all mammalian kidneys go through three stages of embryological development. In fact, these stages are not so much steps in the development of one organ as they are three different, successive, fully functional organs, only the last of which survives into adulthood.

The first of these stages, the *pronephros* (pro-NEH-frose), resembles the kidneys of jawless fishes such as the lamprey. Jawless fish are found in the oldest geological formations, and are considered to be the most primitive of the fishes. The second stage, the *mesonephros* (me-zo-NEH-frose), resembles the kidney of the jawed fishes and amphibians. The third and final stage, the *metanephros* (meh-da-NEH-frose), is the adult kidney found in all terrestrial reptiles, birds, and mammals (see Figure 5-13). Apparently, during our embryo

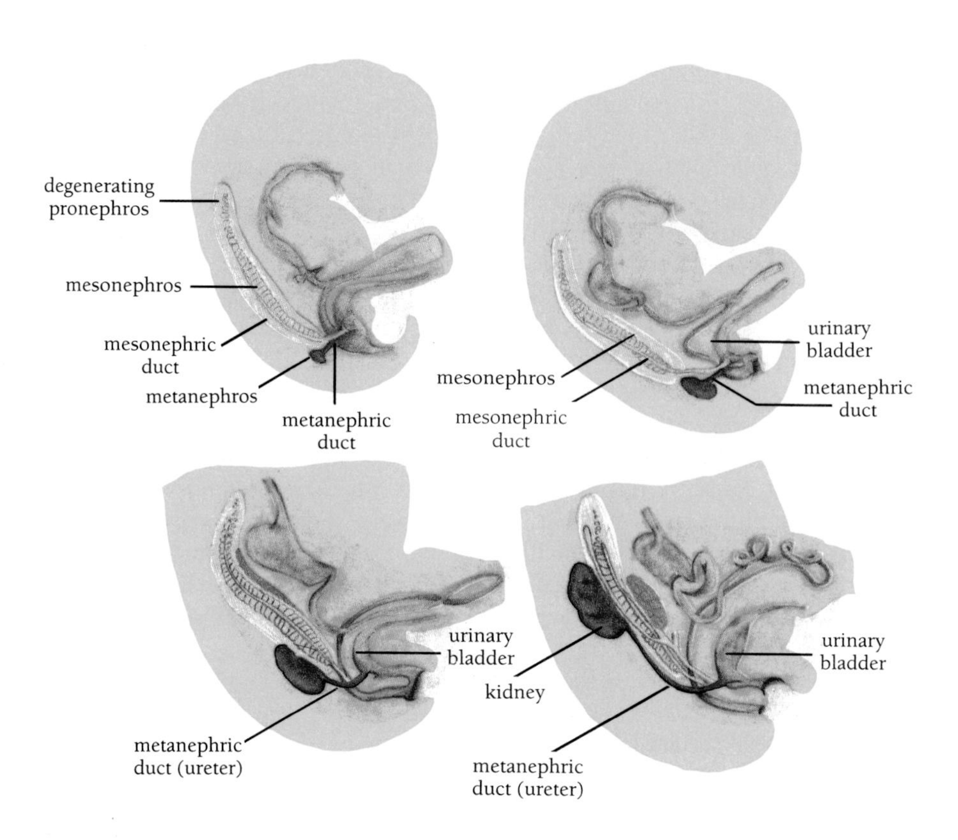

Figure 5-13. ***Human kidney development. The three stages of kidney development in the human embryo are shown, the pronephros, the mesonephros, and the metanephros: organs which progressively replace each other.***

stage we have three kinds of kidneys, each seemingly homologous to an adult kidney in what is considered to be our ancestral stock. Moreover, these kidneys appear during the embryo stage in the same order as they appeared during the supposed evolutionary history of vertebrates. What does this correlation mean?

In order to consider embryonic recapitulation as evidence for evolution, one must make use of the concept of homology. Because of their similarity, the mammalian pronephros and the kidney of the lamprey are assumed to be homologous. Why? Because it is initially believed that macroevolution occurred and, that being assumed, a pronephros so closely resembling the adult lamprey kidney could not have evolved independently; it would have to be homologous.

If living things were designed without macroevolution, the pronephros and lamprey kidney could not be homologous in the Darwinian sense. The concept of recapitulation would also be false because it is based on the assumed truth of macroevolutionary homology. But, you might ask, isn't the observed similarity in embryonic development of the kidney what we would expect if the kidneys were truly homologous? In other words, is not the concept of homology used to predict correctly what we actually found? Is this not one of the tests of a good hypothesis, that it results in correct factual predictions? The answer is, yes; but scientists have recently recognized that embryonic structures once thought functionless actually play important roles in the formation of more advanced structures to follow.[14]

The Skeleton: A Hard Case for Recapitulation

Vertebrates have three basic types of skeletal tissue: *bony, cartilaginous,* and *fibrous.* All three may occur in the same organisms, and all three have a common plan of cellular organization. Most of the volume of each type of tissue, though produced by living cells, is composed of a nonliving substance called the *matrix.* Variation in the characteristics of the matrix accounts for most of the difference among the three kinds of skeletal tissue.

The matrix of fibrous skeletal tissue is composed of strong filaments of the proteins *collagen* (KOL-uh-jin) and *elastin* (see Figure 5-14). This tissue is also suitable for the construction of *tendons* that transmit muscular forces to bones, or of *ligaments* that hold bones together at a moveable joint, or for the deep *dermis* layer of the skin.

The matrix of cartilage is flexible, almost rubbery in texture, because it consists of a mixture of extremely fine fibers and fluid. It occurs in higher vertebrates in place of bone, where bone could be hazardous to one's health. For example, in human beings it occurs in the external ear and the tip of the nose, and where flexibility is needed for the proper function of a part of the skeleton such as the rib cage. It also occurs in moveable joints as an antifriction, load-bearing surface. Because of its fluid component, the cartilage allows nutrients to diffuse readily throughout the matrix. In this way, the living cells found in little cavities distributed through the matrix are able to receive nourishment. This is probably the reason that cartilage in the adult neither needs nor has any blood vessels.

Bone, on the other hand, has a heavily calcified matrix (porous structure) which contains, in addition to collagen, a calcium compound, *apatite* (AP-uh-tite). The apatite makes the bone hard and impermeable to the movement of nutrients to the living cells of the bone. How then does bone tissue survive? It is so organized that no part is very far from a blood supply; in fact, bone has one of the highest concentrations of blood vessels of any tissue. Most human bone is configured into tiny, cigar-shaped groupings of cells called *osteons* (see Figure 5-15). Each osteon is composed of concentric layers of bone cells wrapped around a capillary. In order for material to be able to freely pass from capillary to cell and back, and from one cell to another, tiny filaments of cytoplasm stretch from one cell to another through minute passages called *canaliculi* (ka-nal-IK-u-li). Each cell holds hands with another, so to speak, even though they are separated from each other in their own little caverns in the matrix.

Which is more primitive according to Darwinian thought, bone or cartilage? Cartilage is clearly simpler in organization, and is the only vertebrate skeletal tissue that also occurs in invertebrates (such as squids and horseshoe crabs). Moreover, the living vertebrates considered to be the most primitive (sharks

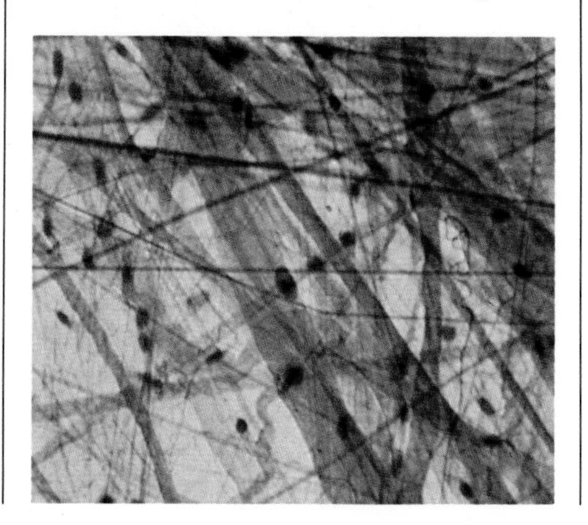

Figure 5-14. ***Collagen fibers.***

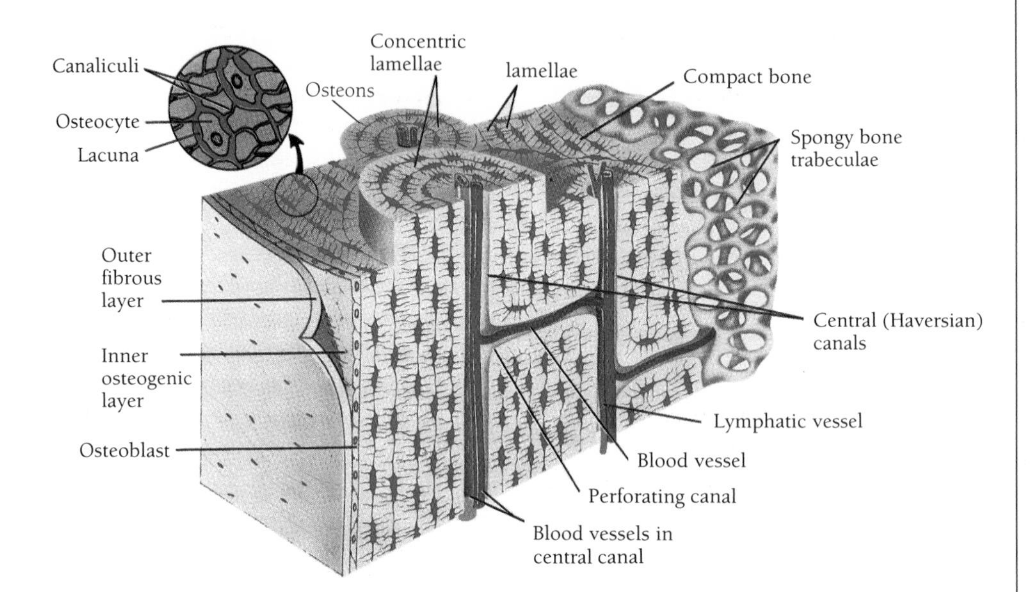

Figure 5-15. ***Bone tissue.***

and modern jawless fish) use only cartilage for skeletal tissue. Surely, then, cartilage is the most primitive tissue. As a final and seemingly decisive argument, consider this: the limb bones of all vertebrates originate as cartilage models and subsequently turn to bone. Much of the original cartilage persists even during infancy and childhood, and a final remnant remains in the adult at bone ends of moveable joints where it serves as nature's "Teflon," an antifriction surface. The cartilaginous "bones" of the fetus are similar in shape and location to the adult cartilaginous bones of our assumed evolutionary ancestors, and so may be considered as homologous to those bones. Since the bones are homologous, it might be predicted that ontogeny will recapitulate phylogeny, and apparently it has. Or has it?

The question of whether the bones of the human fetus really recapitulate the evolutionary development of bone rests on the question of whether cartilage is really more primitive than bone. However, there is one key fact that we have deliberately omitted so far. The jawless fishes found in the supposedly oldest geological formations are heavily armored with a skeletal tissue that is well enough preserved to be examined microscopically. From a study of its structure, the only conclusion that can be reached is that it is bone! Therefore, in vertebrates neither bone nor cartilage is more primitive. They are contemporaries, a state of affairs to be expected if the groups of organisms in which they first appear were brought into existence independently. In fact, "primitive" and "advanced" might well be merely subjective evaluations based on presumed Darwinian ancestry.

Here is a finding that is completely at odds with what might reasonably have been expected on the basis of recapitulation. This illustrates that when we attempt to make predictions about unobserved past events, we are limited to speculation that is not testable. While recapitulation is given less credence today than at one time, some textbooks have not kept pace, and it is still taught. Hopefully, the example of the bone and cartilage will show us the problems that arise when we try to establish a sound view on the idea of recapitulation and the idea of homology which underlies it. Homology appears to be an unreliable concept often leading to difficult and insoluble intellectual puzzles, and at its worst, yielding predictions that are contrary to fact. Since it is not reliable, most of the so-called evidence for macroevolution (and conversely, against intelligent design) obtained from comparative anatomy and embryology is weak and could turn out to be misleading. There is always a high risk in reasoning from an unestablished assumption, no matter how logical that reasoning is.

The Intelligent Design Interpretation of Homology

Based on what we have studied so far, how would you answer the following true or false question: "Design proponents reject the concept of homology." You say, "true!" Sorry, but it's false. They only reject the Darwinian interpretation of homology. Actually, the initial concept of homology was originated by certain German philosophers who were creationists and lived and wrote before Darwin.

As design proponents look at homology today, it appears likely that there must exist a small vocabulary of forms, a vocabulary limited by the functional constraints of a particular set of design objectives, in a particular biological context. If this is correct, this vocabulary of forms or structures would be much like our vocabulary of words. Nothing would compel the choice from this limited set of one form over any other; nothing would prevent the repeated use of a particular form. Hence, choices of individual discretion would remain open to a designing intelligence. Would this not relieve the strain in crediting the blind watchmaker with striking coincidences like the parallels between the giant panda bear and the red panda raccoon?

Darwinian scientists have sought to trace evolutionary relationships between organisms based in some principled way on similarities. But clear principles have been elusive. As you have seen in this chapter, the concept is too vague and too intertwined with the concept of analogy to be a reliable guide to evolutionary relationships, or even to tell us whether macroevolution indeed did occur. Might not innumerable intelligent selections from a vocabulary of forms leave in disarray our efforts to trace evolutionary relationships, exactly the state that now exists?

But what would an intelligent design perspective do to the science of taxonomy? Could it proceed? It is probable that taxonomy would not undergo much change, for when Darwinism arose as the dominant view, it did not substantially change the science of taxonomy as originally launched by creationists.

Perhaps now you can see, too, that the existence of homologous structures merely raises questions of relationship, but it cannot answer them. This is why Stephen Gould remarked that homology supports common design as well as it does common ancestry.[15] Both Darwinists and design proponents can explain the existence of homologies within their respective frameworks of interpretation. Because of this, neither side can disprove the other's interpretation of homology, and neither view stands solely on its own interpretation of homology. Hearing both, however, can stimulate thought and thus be an important tool of education. But it remains for other fields such as mutation studies and biochemistry, to give us more rigorous, quantifiable approaches which can help point the way in our quest to know and understand.

Suggested Reading/Resources

The New Biology, by Robert Augros and George Stanciu. Boston: Shambhala, 1987. A good general discussion of biology with emphasis on the role of intelligence.

References

1. P. Forey, 1932. *Problems of Phylogenetic Reconstruction*, eds. K.A. Joysey and A.E. Friday, New York: Academic Press; and C. Patterson, 1981. *Annual Review of Ecology and Systematics* 12,195-223.
2. D.D. Davis, 1964. *The Giant Panda: A Morphological Study of Evolutionary Mechanisms*. Fieldiana, Zool. Mem. 3, Chicago Natural History Museum.
3. R. Morris and D. Morris, 1966. *Men and Pandas*. New York: McGraw-Hill, p. 153.
4. J.R. Hoener and J. Gorman, 1988. *Digging Dinosaurs*. New York: Workman Publishing, pp. 163-165.
5. S.J. Gould. *Natural History*, September 1992, pp. 10-18.
6. G. de Beer, 1971. *Homology, An Unsolved Problem*. London: Oxford University Press, p. 13.
7. Ibid, p. 16.
8. C.M. Goss, ed. 1988. *Gray's Anatomy* 25th Edition, Philadelphia: Lea and Febiger.
9. H. Kawanishi, 1987. *Immunology* 60,19-28; K. Bjerke, P. Brandtzaeg, and T.O. Rognum, 1986. *GUT* 27,667-674.
10. E. Haeckel, 1866. *General Morphology of Organisms*. Berlin: Georg Reimer, 2:300.
11. K.L. Moore, 1989. *Before We Are Born*. Philadelphia: W.B. Saunders, p. 134; E. Beck, D.B. Moffat, and D.P. Davies, 1985. *Human Embryology*. Osney Mead Oxford: Blackwell Scientific Publishers, p. 172.
12. K.S. Thomson. *American Scientist*, May/June 1988, pp. 273-275.
13. R.A. Raff and T.C. Kaufman, 1983. *Embryos, Genes and Evolution, the Developmental-Genetic Basis of Evolutionary Change*. New York: MacMillan, p. 19.
14. Ibid, p. 18.
15. S.J. Gould. *Natural History*, January 1987, p. 14.

EXCURSION CHAPTER 6

Biochemical Similarities

Introduction

Various types of living organisms that differ dramatically from each other still bear striking resemblances. For example, most land-dwelling vertebrates have a head, body, and four limbs. The skeletons of these animals have a basic structural plan in common. Similarly, the different groups of vascular plants share many features: roots, stems, special vessels for the transport of water and nutrients, and mechanisms for the conservation of water.

Yet even more striking similarities among living organisms occur at the cellular and molecular level. All living organisms, from bacteria to giant redwood trees, are composed of one or more cells. Some cells, such as bacteria, are relatively small and lack a membrane-enclosed nucleus. They also lack the small functional structures called organelles. This group represents a main division of cells, the *prokaryotic* cells. Plant and animal cells are very similar, having the same cellular structures (an enclosed nucleus containing most of the DNA, mitochondria, lysosomes, and ribosomes), cytoplasm, and a cell membrane (see Figure 6-1). Plant cells, however, are surrounded by a rigid cellulose wall, contain chloroplasts that convert solar energy into stored chemical energy, and usually have a large fluid-filled vacuole. Moreover, both plant and animal cells reproduce by *mitosis* or *meiosis*. As you can see,

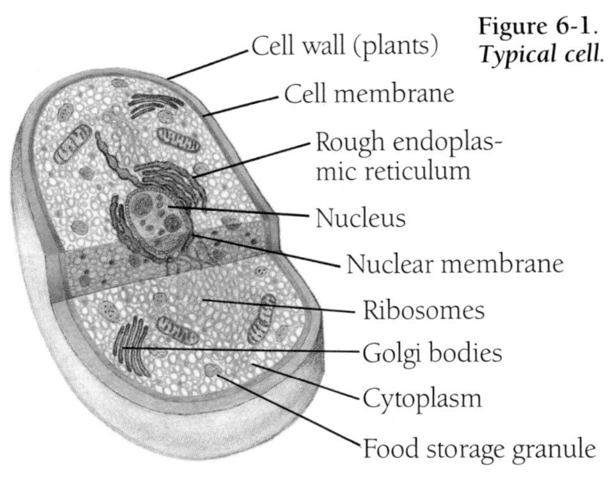

Figure 6-1. ***Typical cell.***

organisms may differ greatly in their general appearance, and yet be strikingly similar at the cellular level.

All living cells contain the same four classes of large biological macromolecules: 1) nucleic acids (DNA and RNA), 2) proteins (for example, oxygen carriers, structural proteins, and enzymes), 3) polysaccharides (starch in plants, glycogen in animals), and 4) lipids (fats). Not surprisingly, these substances perform the same functions in all living things. Genetic information is stored in DNA molecules and transcribed onto RNA molecules; enzymes catalyze the chemical reactions of life; polysaccharides and lipids store chemical energy; and lipids are especially important in membranes.

Another chemical similarity in all living things is found among the smaller molecules (*biomonomers*) of which the large biological molecules are composed. The same 20 amino acids make up the proteins of all organisms. Glycerol and fatty acids are components of lipids; sugars, especially glucose, are the building blocks of polysaccharides; and all DNA and RNA molecules are composed of purines and pyrimidines.

The Structures of Molecules

Protein molecules are long and chain-like in their linear structure, with each amino acid occupying a position in the chain by linking with the amino acids next to it. Most proteins contain from 100 to 500 amino acids in one or more chains of this linear arrangement. The position of certain amino acids in the chain is very important, since the sequence determines how the chain will fold up and also how the protein will function. If we could magnify a protein one million times, it would look like a tiny snarl of thread. Most protein molecules, however, are enormously complex. When folded, each one takes on a characteristic three dimensional surface geometry, with its own precise contours. These precise contours are indispensable if it is to function properly. Since DNA sequences are the source of this sequence information in proteins, they contain the same precise information. The molecular model of horse-heart cytochrome *c* in Figure 6-2 shows some of the complexity of even small proteins.

DNA molecules are much longer than proteins, and instead of individual chains, two strands, running in opposite directions (anti-parallel), spiral round each other to form the famous double helix. As was brought out in Chapter 2, a single strand of human DNA is about 50 million times longer than it is wide!

Common Pathways, Common Code

Not only do all living cells utilize the same basic kinds of chemicals, they also share many of the same basic pathways of chemical reactions. For example, *glycolysis* (gly-KOL-uh-sis), the conversion of glucose into pyruvic acid, occurs in every living cell. This chemical reaction consists of 10 specific steps, each catalyzed by a specific enzyme, and the entire sequence is almost identical in every species. The system of energy-releasing reactions known as the *Krebs cycle* is also virtually identical in all eukaryotic organisms. All oxygen-using (aerobic) eukaryotic organisms have the same complex molecular machinery in their mitochondria, where the Krebs cycle and other energy-releasing reactions take place.

The Meaning of Homology on the Cellular Level

Darwinists take the widespread occurrence of fundamental chemical building blocks, processes, and organization as evidence of the common ancestry of life through macroevolution. Doesn't it make sense, say Darwinists, that the best adapted fundamental elements of ancient life forms would continue to be utilized by living systems, in spite of other changes through time?

One response of design proponents is to cite the importance of the food chain. Food chains provide the support that makes possible large and diverse ecosystems. One organism eats another, that one is eaten by a third, and so on. Because organisms are able to store energy, numerous others can be supported as that energy is passed on. Design proponents reason that common building blocks are a matter of efficiency. If the molecular building blocks used by predators were different from those of their prey, how could they utilize their food? They would have to reduce the molecules taken in beyond the amino acid stage (a process only partly employed by present organisms), and build them back up into usable building blocks

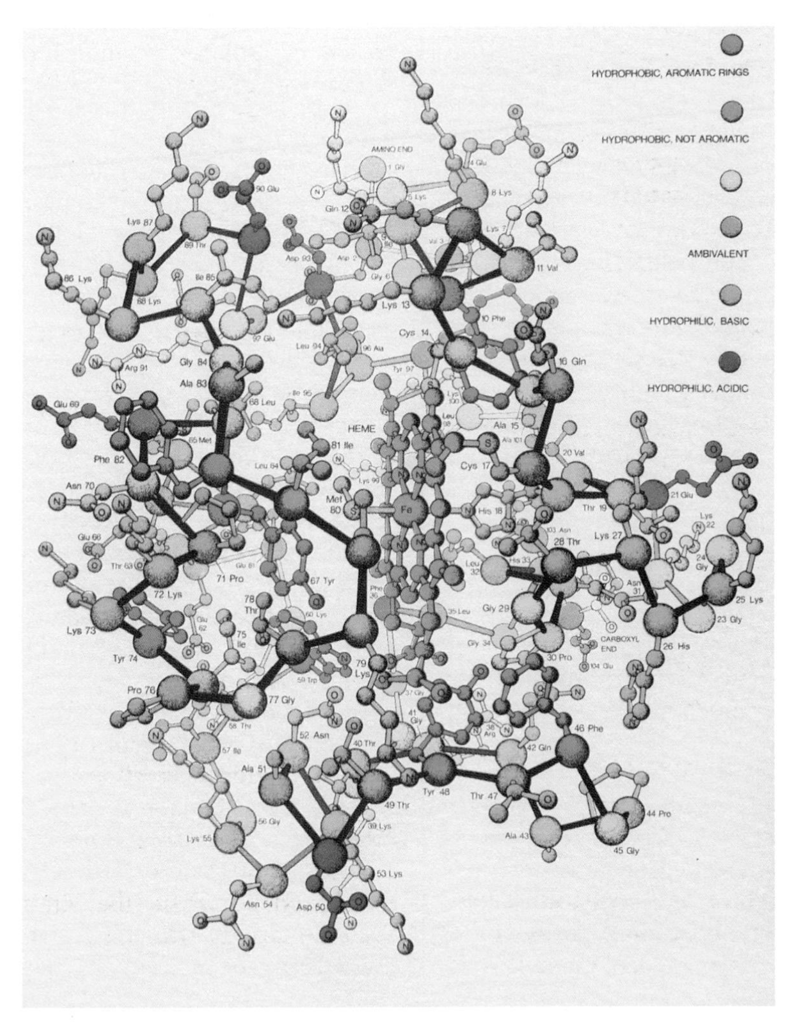

Figure 6-2. A molecular model of horseheart cytochrome c in an oxidized state, showing the side chains.

by a complex series of chemical reactions. This process would require a great additional expenditure of energy. As a result, food chains would be cut short, affecting the entire world of living organisms.

Similarities in metabolic pathways illustrate that common physical and chemical requirements must be met by all living organisms. An intelligent agent, say many scientists, could certainly employ common means, like metabolic pathways, to satisfy such universal requirements. This is no less likely or reasonable than common ancestry.

According to design proponents, structural features and functional processes are held in common across a wide spectrum of organisms because they share a common designer. It should be obvious that the various similarities above do not indicate either descent or design. Arguments based on either common ancestry or common design therefore, reach "compelling" proportions only when they are isolated from their logical alternative. Considering both the Darwinian and design interpretations of the data from morphological similarities shows that we must look to other factors to resolve the case on origins.

Similarities in Molecular Sequences

Darwinian scientists have also pointed to a second, more sophisticated category of similarities between organisms, similarities that are much more detailed and specific.

Since macroevolution, the argument runs, involves changes in the genetic material of species, detailed comparison of genes (DNA) and gene products (proteins) should reveal evolutionary relationships among organisms. For example, comparison of the DNA of humans with the DNA of the living great apes should reveal how closely related the various types of apes are to humans and to each other. Further comparison of the proteins of these species should provide additional information concerning relationships. Theoretically, this approach could be applied to all types of organisms. The only requirements are 1) that we have a way of extracting and purifying DNA and proteins from various species, and 2) that we have methods of determining sequences of genes and proteins with common function in different species. Then the results of comparison should be a molecular "tree of life." Such trees are intended to show the evolutionary relationships between all plants and animals included in the study.

It is easy to understand why Darwinists believe that a study of DNA and protein similarities in different organisms will give a more complete and accurate picture of their evolutionary relationships than a study of the homologous structures of their anatomies. In order for one type of species to change into a new type of species, significant changes would have to occur in some portions of the DNA and, as a consequence, in the proteins for which they code. By comparing the protein and DNA sequences of one species with those of another from a class or order ancestral to the first, it is hoped that the degree of relationship can be measured.

A similiar comparative study of anatomical traits cannot be expected to produce a high degree of accuracy. Why not? First, because similarities in anatomical features are usually approximate, and not easy to measure. Also, environment affects anatomical traits in a given organism, bringing about more variation than is found in their underlying genes. For example, the size and weight of individual organisms can vary depending on diet and state of health, even without any differences in their DNA, as is often seen in identical twins.

Experimental techniques and equipment have become available, allowing the amino acid by amino acid comparison of proteins, such as myoglobin, hemoglobin, and cytochrome *c* from various species of organisms. The amino acid sequences of these and other proteins have been determined, and the results catalogued in the Dayhoff *Atlas of Protein Sequence and Structure*.[1] Darwinists view this ever-increasing data base and comparable information about DNA base sequences as essential information for unraveling the evolutionary history of life.

How Molecular Sequences Are Compared

As long as they are found in corresponding molecules, we can compare such amino acid sequences for diverse species and obtain a measure of their divergence. The Dayhoff *Atlas* uses a table or matrix to compare corresponding molecules from large numbers of diverse species. A convenient way to present these comparisons is by the percent of variation between the corresponding sequences. Table 1 in the Overview Chapter compares the cytochrome *c* molecules of 17 organisms, as drawn from a larger table in "Dayhoff."

The basis for such comparisons is illustrated in Figure 6-3, which has three sets of sequences each containing ten positions (in a protein, each position would be occupied by one amino acid). Notice that the first two sets are identical in all but two positions (numbered 1 and 2). Thus their divergence is 20%. If they differed in three positions, their divergence would be 30%. When comparing any two seqences, we call this the *percent sequence divergence.*

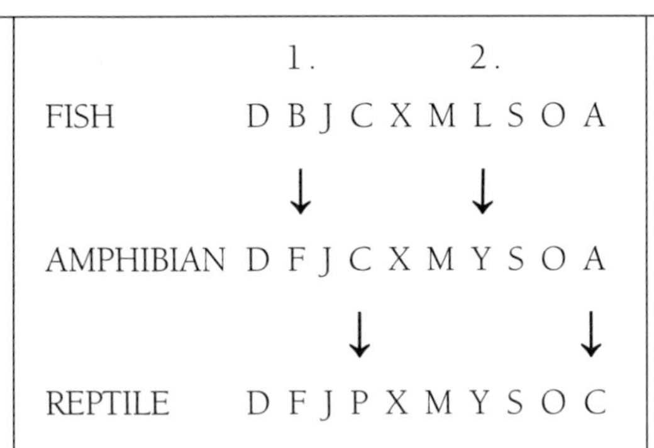

Figure 6-3. ***Three 10-position sequences. Sequences 1 and 2 provide an example of how percent divergence numbers are obtained. The arrows show how changes might appear between sequences. If such changes occurred in the molecular sequences of organisms, they could help determine their relatedness.***

Look again at Figure 6-3. Notice the changes at given positions, this time indicated by arrows. By tracing these changes, you can see that the source of sequence 3 appears to be sequence 2, and the source of sequence 2, sequence 1. On this count, sequence 1 is the "most primitive," sequence 3 is the "most advanced," and sequence 2 is intermediate. This same principle can be extended to much longer series, to search out the evolutionary patterns from most primitive to most advanced.

Look at Figure 6-4, which compares the cytochrome *c* of a dogfish to the cytochrome *c* of six vertebrate classes supposedly widely separated in evolutionary origins over time.

Instead of less divergence from the dogfish for "less advanced" vertebrates and more for "more advanced" vertebrates, the percent of divergence is almost exactly the same for each vertebrate species! When we get down to the business of trying to establish an evolutionary series of sequences, we cannot find the linear, primitive-to-advanced arrangement we had expected. In fact, instead of a progression of increasing divergence, each vertebrate sequence is equally isolated from the cytochrome sequence for the dogfish.

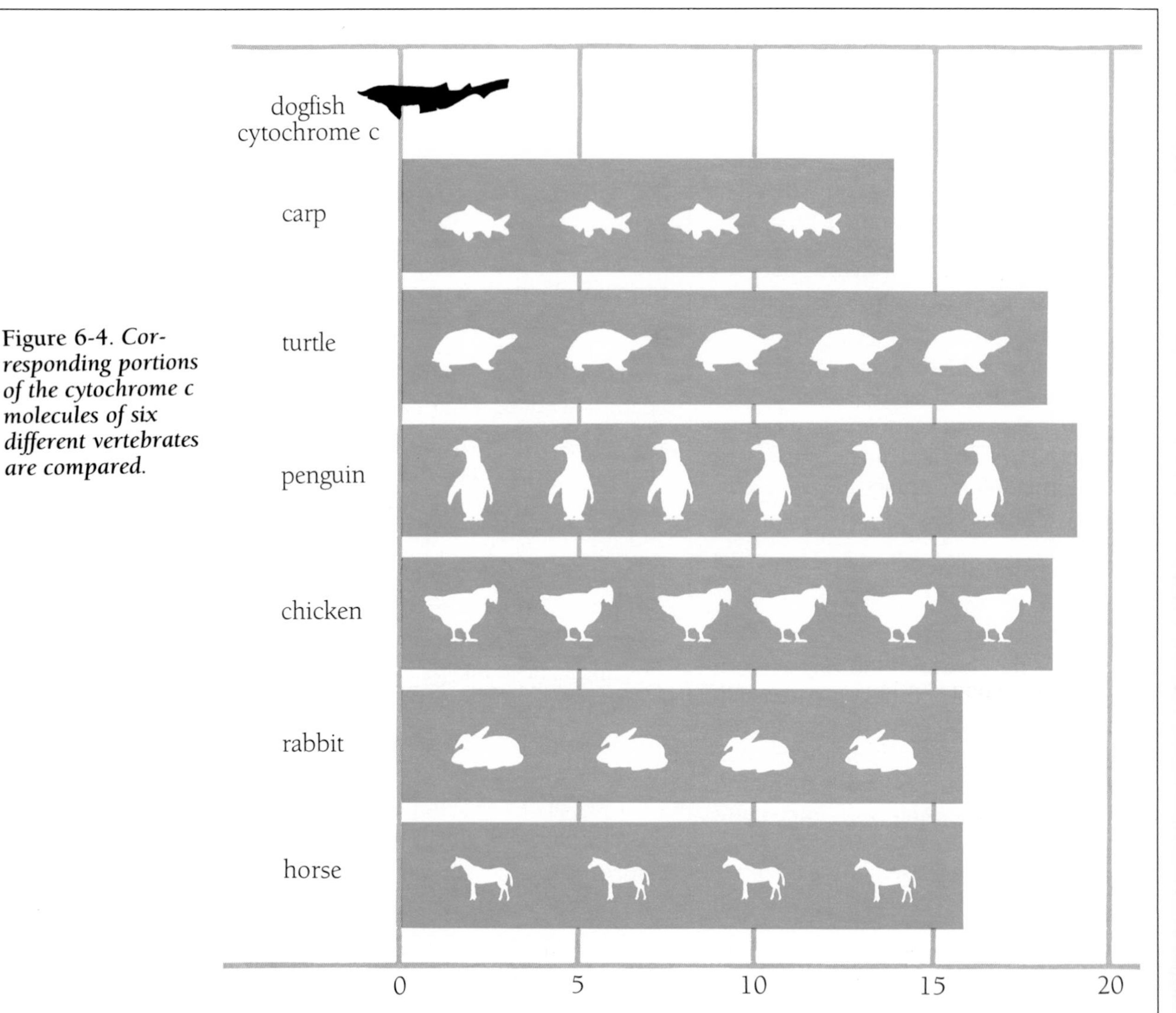

Figure 6-4. ***Corresponding portions of the cytochrome c molecules of six different vertebrates are compared.***

In this and countless other comparisons, it has proved impossible to arrange protein sequences in a macroevolutionary series corresponding to the expected transitions from fish→amphibian→reptile→mammal. There is no hint of an intermediate in these data. All are virtually equidistant from the dogfish. This is truly amazing because amphibia are usually considered intermediate between fish and mammals.

What is found in the molecular data, instead, is astounding by any measure. Rather than showing sequences that are intermediate between subclasses, every sequence can be clearly identified as to which subclass it belongs, with little shading or ambiguity beyond the species level. Furthermore, all sequences of a particular subclass are equally distant from the sequences in a more distant group, or an "outgroup."

This is an unexpected parallel to the clustering pattern that has become quite familiar to us from both living and fossil organisms.

A Biochemical System at Work

It is fascinating to see how biochemicals operate in living systems. Biochemical processes underlie the various organismal structures and systems we have seen throughout the book. So let's look at an example of proteins operating together in the environment of cell and tissue, to see how they work.

When a container of liquid, like a can of soda, springs a leak, the fluid quickly drains out. However, when a person suffers a cut it ordinarily bleeds for only a short time before a clot forms to stop the bleeding. Soon the clot hardens and eventually the cut heals over. Blood clot formation seems so familiar to us that most people don't give it much thought. However, biochemical investigation has shown that blood clotting is a very complex, intricately-woven system containing a score of interdependent protein parts. [2] The absence or defective operation of any of several of these components will cause the system to fail, and blood will not clot at the proper time or at the proper place. In this respect, clotting is like many other biological systems. Continuing research is showing that most biochemical systems are composed of interactive parts in which many different components have to operate simultaneously for a functioning system to occur, much like a car motor requires the simultaneous operation of spark plugs, pistons, a radiator, a fan belt, and more, if it is to run. As we shall see, such interactive systems, as illustrated here by the mechanism for blood clotting, are very strong arguments for intelligent design and are virtually impossible to explain in terms of Darwinian evolution.

The Mechanism of Blood Clotting

Blood clotting has to work within very narrow restrictions. When a cut occurs in an organism with a pressurized blood circulation system like ours, a clot must form quickly or the organism will bleed to death. On the other hand, if clots occur at times or places other than the site of an injury, they may block blood circulation, as they do in heart attacks and strokes. Furthermore, when a cut occurs the clot has to stop the bleeding all along the length of the cut, sealing it completely. But blood clotting must also be confined to the area of the cut or the entire blood system of the animal could solidify, killing it. Therefore the clotting of blood must be tightly controlled to make sure the clot forms when it needs to and not otherwise. To see how this control is exerted, let's explore what happens when a person gets cut.

About two to three percent of the protein in blood plasma consists of a protein called fibrinogen. Normally, the fibrinogen is dissolved in the plasma like salt is dissolved in ocean water. But when a cut occurs, another protein called thrombin slices off a piece of the fibrinogen protein. The smaller protein, now called fibrin, is sticky, and large numbers of fibrin proteins aggregate with each other. They don't aggregate randomly, however. Because of the shape of the fibrin molecule, long threads form, cross over each other, and make a protein meshwork that entraps blood cells (see Figure 6-5). This forms the initial clot.

But we have a chicken-and-egg problem here: thrombin activates fibrinogen to form the sticky fibrin; yet what tells thrombin to begin its work? It turns out that, like fibrin, thrombin initially exists in an inactive form, called prothrombin.

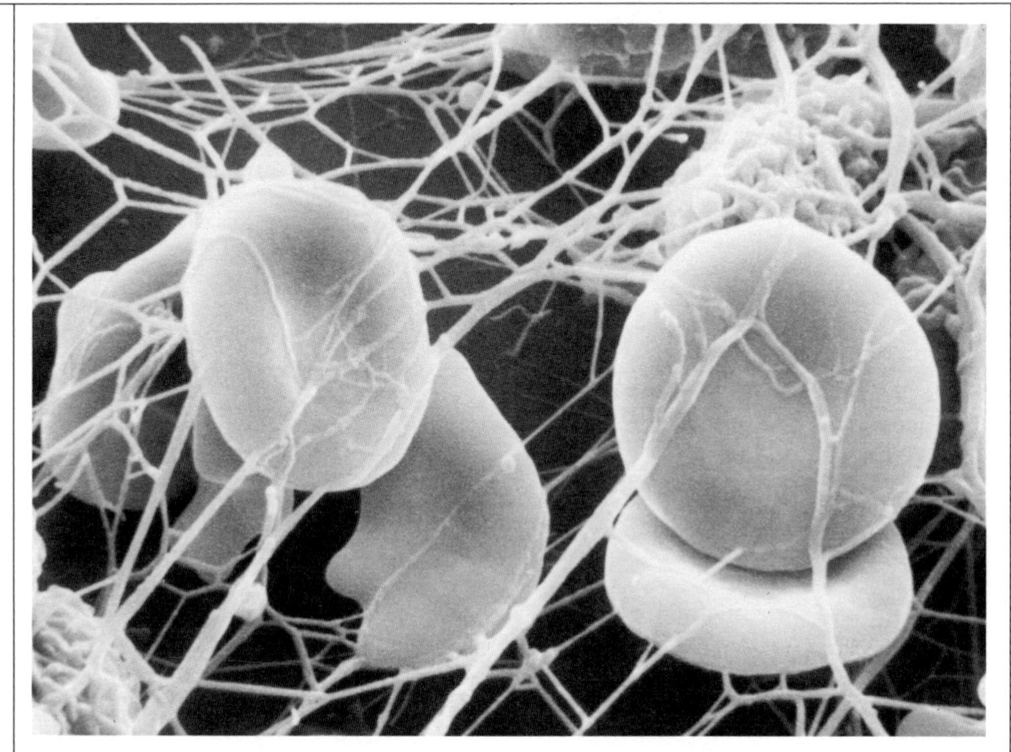

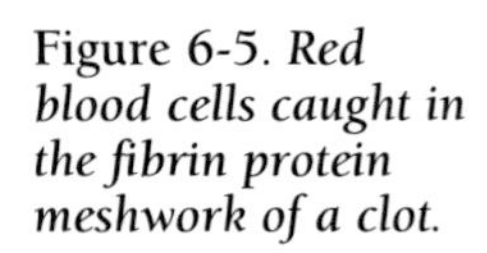

Figure 6-5. *Red blood cells caught in the fibrin protein meshwork of a clot.*

Prothrombin cannot cleave fibrinogen to fibrin, but when two other proteins, called *Stuart factor* and *proaccelerin* (the active forms of the proteins are indicated by italics), cut off a small portion of the prothrombin molecule to make thrombin, then thrombin can start to work (see Figure 6-6).

Yet as it is initially made by the cell, prothrombin cannot be turned into thrombin, even in the presence of *Stuart factor* and *proaccelerin*. It must first be modified by having several of its amino acid **residues**, called glutamate (Glu) residues, changed to γ-carboxyglutamate (Gla) residues. This is necessary to allow prothrombin to bind calcium and then stick to the inside, exposed surface of an injured cell. Only the intact, modified, calcium-prothrombin complex bound to the injured cell can be cleaved by *Stuart factor* and *proaccelerin* to give thrombin. Now, it has been shown that the conversion of Glu to Gla requires an additional component: vitamin K ("K" for "coagulation," since vitamin "C" was already taken). Vitamin K is not a protein, but it is necessary for some proteins to do their jobs. Specifically, the protein which modifies prothrombin to a form containing Gla residues needs vitamin K to enable it to work. When animals ingest poisons that prevent the vitamin from doing its job, prothrombin is neither modified nor cleaved, and the poisoned animals bleed to death.

In spite of all this "molecular teamwork," we are still many steps away from a blood clot. Now we have to ask what activates Stuart factor. It turns out that it can be activated by two different routes, called the "intrinsic pathway" and the "extrinsic pathway." (In the intrinsic pathway all the proteins required for clotting are contained in the blood plasma; in the extrinsic pathway some proteins required for clotting occur on cells.) The intrinsic pathway begins with the activation of

Hageman factor at the top left of Figure 6-6; the extrinisic pathway begins with the activation of proconvertin at the far right of the same figure. Notice the number of components in each pathway. Notice also that only when an injury brings tissue into contact with blood will the extrinsic pathway be initiated. In either pathway, we repeatedly see one component activating another. Why is blood clotting so complex? Because the clot has to form rapidly to stop the flow of blood and a "cascade" of components provides the most rapid means of bringing this about.

The intrinsic and extrinsic pathways cross over at several points. Hageman factor, activated by the intrinsic pathway, can activate proconvertin of the extrinsic pathway. *Proconvertin* can then feed back

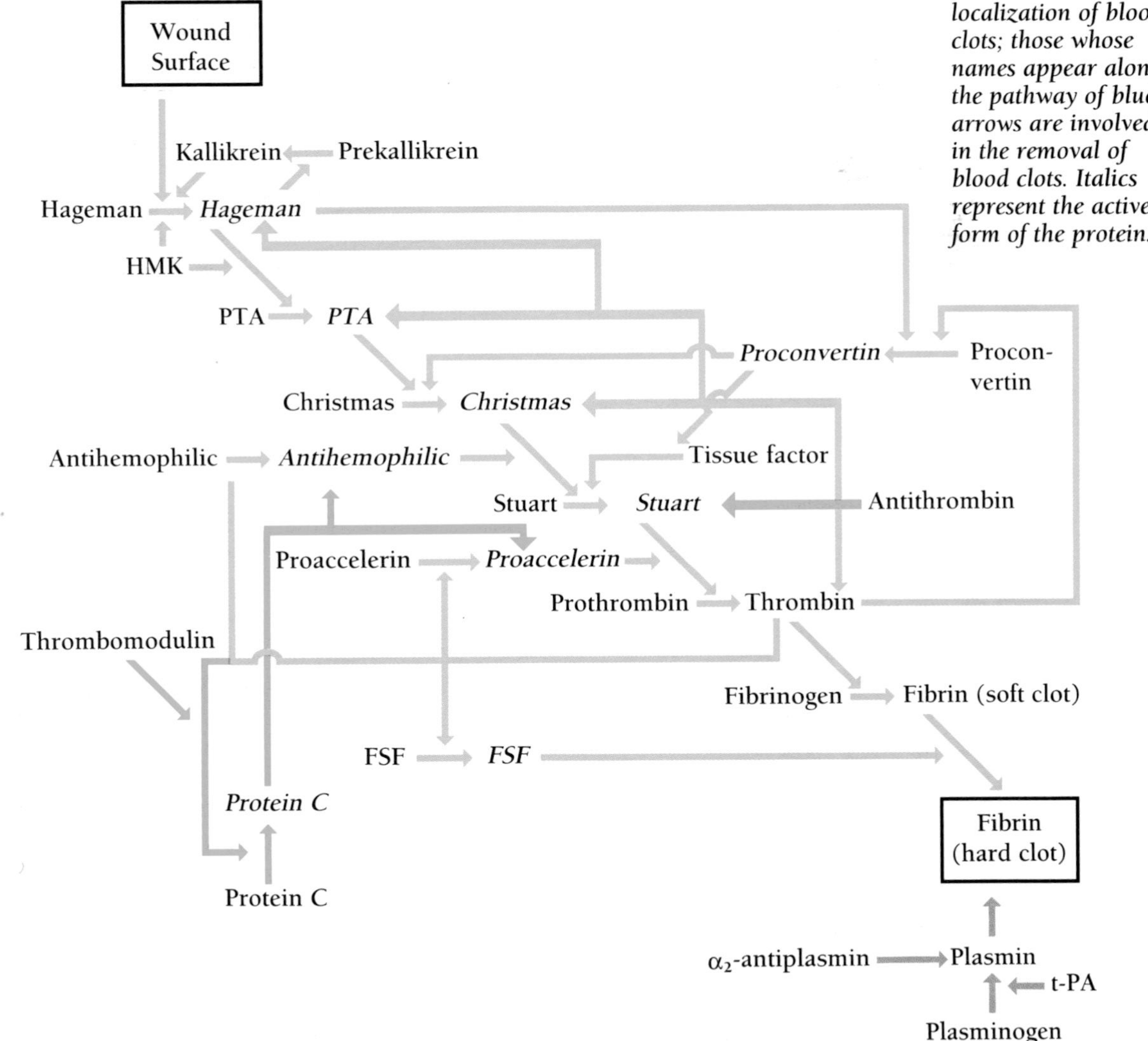

Figure 6-6. *Outline of the blood clotting system. Proteins whose names appear along the pathway of green arrows are involved in promoting clot formation; proteins whose names appear along the pathway of orange arrows are involved in the prevention and localization of blood clots; those whose names appear along the pathway of blue arrows are involved in the removal of blood clots. Italics represent the active form of the protein.*

into the intrinsic pathway to help *PTA* activate Christmas factor. Thrombin can also trigger both branches of the clotting cascade.

Localization of the Blood Clot, Strengthening, and Dissolution

As seen by the previous discussion, the blood clotting cascade is complex and intricate. But once clotting has begun, what keeps it from continuing until all the blood in the organism has solidified? Here, too, the teamwork of several proteins is involved. Figure 6-6 shows two separate mechanisms that restrict the formation of the clot to the injured tissue.

Still other protein machinery is on standby, ready to begin the process of blood clot removal. This process, which involves the breakdown and removal of the clot, must not begin too early, however, or it will prevent healing. It therefore awaits the proper biochemical signal to begin.

The success of the blood clotting pathway depends critically on the rate at which the different reactions occur. An organism would not long survive if thrombin activated protein C at a significantly faster rate than proaccelerin, or if antithrombin inactivated *Stuart factor* as fast as it was formed. Thus, blood clotting not only depends on the occurrence of the reactions outlined in Figure 6-6, but also on the control of their timing.

The formation, limitation, strengthening, and removal of a blood clot is an integrated biological system, and problems with single components can cause the system to fail. In this characteristic, it is like the car engine mentioned earlier, which fails to work if the fan belt is missing, or the distributor cap, or the spark plugs; and is severely hampered even by a flat tire. The lack of certain blood clotting factors, or the production of defective factors, often results in serious health problems or death. The most common form of **hemophilia** occurs because of a lack of *anti-hemophilic factor*, which helps *Christmas factor* in the conversion of Stuart factor to its active form. Lack of Christmas factor is the second most common form of hemophilia. Severe health problems or even death can also result if defects occur in other proteins of the clotting pathway.

A Characteristic of Intelligent Design

Why is the blood clotting system an example of intelligent design? The ordering of independent pieces into a coherent whole to accomplish a purpose which is beyond any single component of the system is characteristic of intelligence. Earlier in the book it was pointed out that we recognize an intelligence behind even a short, simple message like "John loves Mary." The sentence is constructed from several independent pieces (the letters) to accomplish a larger purpose (the message). In a similar manner, the purpose of the blood clotting system is to quickly form a clot at the sight of a cut while avoiding clot formation at inappropriate times or places.

However, when isolated from its "molecular team," each of the many components of the system can accomplish nothing constructive, like a steering wheel that is not connected to a car. Yet thousands of other proteins in the blood plasma and cell cytoplasm are exposed to these blood-clotting components without effect. But when activated in the presence of the entire suite of blood-clotting molecules, each one performs its specialized task on the next with precision. When the system is lacking just one of the components, such as antihemophilic factor, severe health problems often result. Only when all the components of the system are present and in good working order does the system function properly.

Creeping toward Clotting

Why is the blood clotting system incompatible with a nonintelligent evolutionary view of nature? Macroevolution means a change from a simpler to a more complex state. Let us try to envision such a change for blood clotting. Assume that we initially start with an organism that contains just a primitive version of thrombin and fibrinogen. The thrombin would immediately cut all the fibrin, causing a massive clot and the speedy death of the organism. Suppose instead we started with fibrinogen and prothrombin. In this case there is nothing to initiate clotting when a cut occurs and the organism would bleed to death. We may try many smaller sets of components to get started—fibrinogen, prothrombin, activated Stuart factor and proaccelerin, or inactive *Stuart factor* or *proaccelerin*, or fibrinogen plus an imaginary protein that cleaves fibrinogen to fibrin—death is nearly always the certain result. *In fact, having a primitive, poorly controlled clotting system would probably be more dangerous to an animal, and therefore less advantageous, than having no such system at all!* Thus the blood clotting system cannot have emerged piecemeal. Like a car or a sentence, it requires the cooperative interaction of pre-existing components to work.

How do Darwinists explain the origin of the blood clotting system? They don't, at least not in any detailed, step-by-step fashion. It is important to realize that no one has ever offered a credible hypothesis to explain how the blood clotting system could have started and subsequently evolved. Nor have they explained how a single protein molecule could be formed by gradual chance events. Instead, Darwinists are content to point to resemblances in protein sequences, as discussed earlier and simply assume that such resemblances mean that gradual evolution somehow occurred. This is the same defense that Darwinists offer at the whole-organism level: similarities among different types of animals are assumed to support the occurrence of Darwinian evolution, but no detailed, testable explanation is offered for how the many integrated biological systems may have arisen. No answer has been forthcoming to the person who asks for details.

We have closely examined the blood clotting system in this chapter and shown that it exhibits characteristics strongly suggestive of intelligent design. It is not unique in this respect: virtually all biochemical systems, large and small, exhibit coherent integration of distinct parts to give a whole entity with a separate purpose. This includes photosynthesis, cell replication, carbohydrate, protein, and lipid metabolism, vision, the immune system, and numerous others. Like a car engine, biological systems can only work after they have been assembled by someone who knows what the final result will be.

There is both elegance and astonishing complexity in even one such biochemical system. Each of these specialized functions traces back to the molecule's amino acid sequence. The amino acids in the suite of blood clotting proteins vary from roughly 300 to 3,000 residues. Some of them share discrete regions of their sequences with some others. Does this mean that they derived from one another? It may, but consider that even if this were the case, *all of the proteins had to be present simultaneously for the blood clotting system to function.*

Nature's Search for a Protein

When we try to compare sequences from one functional class of proteins to another, we meet an even greater challenge for Darwinism. We have already seen that proteins function by folding into precise, stable shapes, and that the detailed contours of these shapes are determined by the specific sequences of the amino acids in their chains. If every position in the chain must have a certain amino acid, the odds of coming to such a specific sequence for a 100-unit protein by an undirected search—or nonintelligent means—would be $1/10^{130}$. This is virtually the same as saying there is no chance that it could have happened without intelligence.

But amino acids at many positions can be substituted by others without sacrificing the function of the molecule. So the question is, How specific must these sequences be? How much variation can be tolerated, at how many positions? For if they have always had to be highly specific, there is no physical or chemical process that could account for their origins. But might there be "stepping-stones" so to speak, by which very different protein sequences with other functions from ancient organisms could have eventually evolved into all of the present day proteins by sequential amino acid substitutions?

To make a specific protein in this way, many sequences, each progressively closer to the target sequence, must have developed, and they must also have all been functional, meaning that they, too, must have folded into shapes with at least some minimal degree of function. Eventually, such intermediate sequences would also have to reach sufficient approximation to the protein sequence to serve, however imperfectly, the function it does today. Once a protein becomes useful for some specific function, the leeway for substitute amino acids at various positions in its chain (while maintaining its function) may be called its *ambiguity.*

In 1978, one of the world's leading information scientists, Hubert Yockey, published calculations that allowed for the ambiguity in a certain number of positions of a 100-amino acid protein.[3] Through these calculations, Yockey showed that this ambiguity provided a dramatic increase in the likelihood that an undirected search for a given functioning protein would be successful. According to Yockey's figures, the odds against an undirected search arriving at such a protein were dropped from $1/10^{130}$ to $1/10^{65}$.

Yet this was still highly unfavorable to Darwinism. 10^{65} is a very large number. For example, it has been calculated that there are approximately 10^{65} atoms in a galaxy. The odds against success in the undirected search for this protein are roughly the same as marking a single atom with an "X", returning it at randon to some location of the Milky Way, and then having a second person (one with access to the entire Milky Way but no knowledge of the whereabouts of the marked atom) picking it out on the first random draw!

But even though Yockey's calculations seemed to agree with what was known from studies of the actual variety of cytochrome *c* molecules, how could we know for sure that his theoretical calculations were correct? Indeed, other scientists, such as Ken Dill at the University of California at San Francisco, did calculations based on different assumptions to answer the same question, and the result of their calculations came out much lower.[4]

To resolve this dilemma, hard experimental results would be required. In the 1980s, techniques were developed to actually tailor-make protein sequences by cutting up and splicing the corresponding strands of DNA, making whatever alterations were desired in its reassembly. Utilizing these techniques in the biological laboratories at MIT, research has recently been done under the direction of Robert Sauer that gives us solid empirical data for how much variation can be tolerated.[5] Investigating two different proteins of approximately the same size as cytochrome *c* molecules, Sauer and fellow researchers systematically substituted the 20 amino acids in living things, three at a time, in each position of these proteins. The resulting DNA strand for each variation was then reinserted into bacteria, which would have used the resulting proteins had they been functional. On the other hand, the bacteria would have destroyed any nonfunctional sequences. Through this means, Sauer's team was able to test every possible amino acid at every position in the chain. They found that up to 15 of the 20 amino acids worked equally well in some positions, while others could tolerate one, two, or a few substitutions, and still others, none whatever.

Later, Sauer used simple calculations from this empirical data, to return to the question raised in Yockey's article. These calculations showed that the odds of finding a specific folded protein are about $1/10^{65}$, a striking confirmation of Yockey's calculations. It means that all proteins that have been examined to date, either experimentally or by comparison of analogous sequences from different species, have been seen to be surrounded by an almost infinitely wide chasm of unfolded, nonfunctional, useless protein sequences. There are in fact, no "stepping-stones"! In other words, an undirected search will not hit upon any of the end protein sequences sought in the time allowed by the age of the universe. The various functional classes of proteins apparently are so highly isolated, they could not have arisen from one another.

Conclusion

Throughout the six previous chapters we have sampled representative data related to origins from different areas of biology. We saw mounting problems in the assumptions underlying chemical evolution accounts of the origin of life. We observed how biological structures exhibit the characteristics of manufactured things. We became aware of the enormous difficulties for the chance production of functional information in the message text of DNA. The demands of this information cast their shadow on the macroevolutionary hypothesis that mutations might somehow bring about the elaborate and specific complexity of biological structures. Moreover, we have noted that these observations are consistent with the results of a wide variety of breeding experiments. Despite vigorous efforts, no new levels of complexity have been achieved in experimental breeds. Instead, changes have been quite limited, and highly bred organisms have often suffered a loss of genetic information.

The fossil record accords equally well with this picture, offering us not graded series of fossil organisms over time, but a record of organisms that cluster in taxonomic groups of similar organisms, distant from other high level taxa, and remaining remarkably stable through time. Finally, we turned from the fossil record with its ambiguities to the area of biochemistry with its promising capacity to measure differences between organisms.

Darwinists have held high expectations that biochemistry would provide evidence of gradual change between taxanomic groups. Such evidence would be doubly important, since it could show a material basis for progressive sequence—descent with modification—and thus offset the failure of the fossil record to do so, as we saw in Chapter 4. However, biochemistry has not provided this kind of evidence.

Over the last quarter century, we have learned many of the informational sequences that play critical roles in the functioning of proteins and DNA, and thus of cells. While we have much to learn about what those roles are, we have made a good start. As we witnessed in the the blood-clotting mechanism, proteins are integrated into systems where they can serve ends beyond those of any individual protein or other component.

When it comes to the matter of comparison, we found that here, too, organisms are distinct. Although the similarities among them carry over from the anatomical level to the level of biochemistry, they do not exhibit a sequential pattern in which one form leads into the next. Instead, there are divisions between the various groups far too uniform to have predicted.

There is now good empirical evidence that classes of proteins with varied functions in living organisms are separated by distances (in their linear sequences) so vast that it is mind-boggling. Because sequence changes in such informational molecules are considered to be the foundation for the changes pictured by neo-Darwinism at the cellular level as well as at the level of traits, structures, and organs, this finding goes to the heart of the neo-Darwinian story.

Any view or theory of origins must be held in spite of unsolved problems; proponents of both design and unplanned descent acknowledge this. Such uncertainties are part of the healthy dynamic that drives science. However, without exaggeration, there is impressive and consistent evidence, from each area we have studied, for the view that living things are the product of intelligent design.

Suggested Reading/Resources

Evolution: A Theory in Crisis, by Michael Denton. Bethesda, MD: Adler & Adler, 1986. A penetrating study of classification and biochemistry.

References

1. M.D. Dayhoff, 1972. *Atlas of Protein Sequence and Structure*. Silver Spring, MD: National Biomedical Research Foundation.
2. B. Furie and B. Furie, 1988. *Cell* 53,505-518.
3. H.P. Yockey, 1978. *J. Theoret. Biol.* 67,377-398.
4. K.F. Lau and K.A. Dill, 1989. *Macromolecules* 22,3986-3994; H.S. Chan and K.A. Dill, 1990. *Proceedings of the National Academy of Sciences USA* 87,6388-6392.
5. J.U. Bowie and R.T. Sauer, 1989. *Proceedings of the National Academy of Sciences USA* 86, 2152-2156; J.U. Bowie, et al., 1990. *Science* 247,1306-1310; J.F. Reidhaar-Olson and R.T. Sauer, 1990. *Proteins: Structure, Function and Genetics* 7,306-316.

GLOSSARY

Allele—Alternative forms of the same gene locus. An allele, therefore, is one of a pair of genes that occupy corresponding positions on homologous chromosomes and determine alternate expressions of a single trait.

Allopatric speciation—The development of a new species through isolation from its ancestral population by geographical barrier.

Analogy (analogous structure)—A body part similar in function to that of another organism, but only superficially similar in structure, at most. Such similarities are regarded, not as evidence of inheritance from a common ancestor, as in homology, but as evidence only of similar function.

Blind watchmaker—Richard Dawkins' metaphor for nonintelligent, purposeless, natural selection, expressing the idea that characteristics apparent in living things like the intelligent purpose and craftsmanship associated with watchmaking are only illusory.

Bottleneck effect—The reduction of a population's gene pool and the accompanying changes in gene frequency produced when a few members survive the widespread elimination of a species.

Catalytic—The property of increasing the rate of a chemical reaction without being used up in the process.

Chemical evolution—The theory that states life first originated as the result of a protracted, gradual transformation of nonliving matter and energy into a living cell.

Coacervates—An aggregate of colloidal droplets held together by electrostatic charges.

Convergent evolution—The evolution of similar appearances in unrelated species.

Darwinism—The theory that all living things descended from an original common ancestor through natural selection and random variation, without the aid of intelligence or nonmaterial forces.

Design—See Intelligent Design.

Dominance, (dominant)—The phenotypic expression of one allele over another allele when both are present in a heterozygote (thus masking the effect of the second allele).

Evolution, (evolve)—Often defined as mere "change in living things over time," but variously meaning also, descent with modification, and particular mechanisms to account for change such as natural selection and gene mutation.

Fixity—The idea of the inability of species to change.

Founder effect—Reduction of the gene pool and alteration of gene frequencies resulting from the establishment, by two or a few organisms, of a new population isolated from the parental population.

Functional information—Information in the base sequence of a species' DNA that codes for structures capable of biological functions.

Gene pool—The total genetic material in the population of a species at a given point in time.

Genetic drift—Changes in gene frequencies of small populations that result from random mating.

Genome—The total DNA for a given organism.

Genotype—The combination of alleles inherited for a particular trait.

Gradualism—The view that evolution occurred gradually over time, with transitional forms grading finely in a line of descent.

Hemophilia—A genetic disorder of the blood clotting mechanism produced by the absence of either of two proteins essential to the formation of blood clots, which thereby allows uncontrolled bleeding from even minor causes.

Heterozygous—Having a mixed gene pair, in which one allele is dominant and the other recessive.

Homology (homologous structure)—A body part with the same basic structure and embryonic origin as that of another organism, a fact taken to imply the common ancestry of the two.

Hypothesis—In scientific method, 1. An educated guess, 2. A tentative explanation, or 3. A proposition that is to be confirmed by test.

Information theory—A branch of applied mathematics which provides a measure of information in any sequence of symbols.

Intelligent design (cause)—Any theory that attributes an action, function, or the structure of an object to the creative mental capacities of a personal agent. In biology, the theory that biological organisms owe their origin to a preexistent intelligence.

Intermediate—A morphological condition or appearance between two others that are more extreme in their differences; an organism which, regardless of genealogical relationships, holds features or structures of body form in common with any two other organisms which do not hold them in common with each other.

Macroevolution—The hypothesis of large-scale changes, leading to new levels of complexity.

Microevolution—Small-scale genetic changes, observable in organisms.

Mimetic butterflies—A species of butterfly with striking resemblance to another species that can serve as a strategy of defense.

Morphology—The form or structure of an organism.

Natural selection—One genotype leaves more offspring for the next generation than other genotypes in that population. The explanation is that the elimination of less suited organisms and the preservation of more suited ones result from pressures of the environment or competition or both.

Nucleotide—The fundamental structural unit of a nucleic acid, or DNA, made up of a nitrogen-carrying base (purine or pyrimidine), a sugar molecule, and a phosphate group.

Parallel evolution—The development of similarities in separate but related evolutionary lineages through the operation of similar selective factors on both lines.

Polymorphic—Descriptive of a species population with diverse body forms.

Prebiotic—Prior to the existence of biological life.

Proteinoids—Racimic chains of amino acids synthesized by the removal of water, which do not exhibit the specified sequences or alpha bonding characteristic of proteins.

Punctuated equilibrium—The theory that speciation occurs relatively quickly, when spurts of rapid genetic change "punctuate" the "equilibrium" of primarily constant morphology known as stasis over geologic time.

Recessive—A gene which is masked, or not expressed in the presence of its dominant allele.

Reducing atmosphere—An atmosphere devoid of molecular oxygen, but rich in hydrogen available for energy-storing combination with other substances.

Residue—A molecule incorporated in another, larger molecule.

Selection pressure—The intensity with which an environment tends to eliminate an organism, and thus its genes, or to give it an adaptive advantage.

Selective advantage—A genetic advantage of one organism over its competitors that causes it to be favored in survival and reproduction rates over time.

Sibling species—Two or more species that are very close in physical appearance (morphology), but not necessarily genealogically.

Speciation—The development of a new species by reproductive isolation of organisms from their ancestral population, resulting in the loss of interfertility with it.

Spontaneous generation—The unaided development of living organisms from nonliving material or from other organisms.

Stabilizing selection—Natural selection operating to eliminate extreme members of a population, thereby reducing variation in a population.

Stasis—The primarily constant morphology of a species over a long period of geologic time.

Taxon—(plural, taxa) A group or category of biological classification at any level.

Theory—A scientific explanation of well-established observations.

Transitional—An organism that holds, in common with other organisms presumed to be its ancestor and descendent, particular features or structures of body form not held in common by the other organisms themselves.

Vestigial structure (organ)—A body part that has no function, but which is presumed to have been useful in ancestral species.

A Note to Teachers

by Mark D. Hartwig, Ph.D.
and Stephen C. Meyer, Ph.D.

Biological origins can be one of the most captivating subjects in the curriculum. As a biology teacher, you have probably already seen how the topic excites your students. The allure of dinosaurs, trilobites, fossilized plants, and ancient human remains is virtually irresistible to many students. Indeed, many prominent scientists owe their interest in science to an early exposure to this topic.

The subject of origins, however, is not only captivating. It is also controversial. Because it touches on questions of enduring significance, this topic has long been a focal point for vigorous debate—legal and political, as well as intellectual. Teachers often find themselves walking a tightrope, trying to teach good science, while avoiding the censure of parents or administrators.

To complicate things, the cultural conflict has been compounded by controversies within the scientific community itself. Since the 1970s, for example, scientific criticisms of the long-dominant neo-Darwinian theory of evolution (which combines classical Darwinism with Mendelian genetics) have surfaced with increasing regularity.[1] In fact, the situation is such that paleontologist Niles Eldredge was driven to remark:

> If it is true that an influx of doubt and uncertainty actually marks periods of healthy growth in science, then evolutionary biology is flourishing today as it seldom has in the past. For biologists are collectively less agreed upon the details of evolutionary mechanics than they were a scant decade ago.[2]

Moreover, many scientists have advocated fundamental revisions of orthodox evolutionary theory.[3]

Similarly, the standard models explaining chemical evolution—the origin of the first living cell—have taken severe scientific criticism.[4] These criticisms have sparked calls for a radically different approach to explaining the origin of life on earth.

Though many defenders of the orthodox theories remain, some observers now describe these theories as having entered paradigm breakdown[5]—a state where a once-dominant theory encounters conceptual problems or can no longer explain many important data. Science historians Earthy and Collingridge, for example, have described neo-Darwinism as a paradigm that's lost its capacity to solve important scientific problems.[6] They note that both defenders and critics find it hard to agree even about what data are relevant to deciding scientific disagreements. Putting it more bluntly, in 1980 Harvard paleontologist Stephen Jay Gould pronounced the "neo-Darwinian synthesis" to be "effectively dead, despite its persistence as textbook orthodoxy."[7]

In this intellectual and cultural climate, knowing how to teach biological origins can be exceedingly difficult. When respected scientists disagree about which theories are correct, teachers may be forgiven for not knowing which ones to teach.

An Opportunity

Controversy is not all bad, however, for it gives teachers the opportunity to engage their students at a deeper level. Instead of filling young minds with discrete facts and vocabulary lists, teachers can show their students the rough-and-tumble of genuine scientific debate. In this way, students begin to understand how science really works. When they see scientists of equal stature disagreeing over the interpretation of the same data, students learn something about the human dimension of science. They also learn about the distinction between fact and inference—and how background assumptions influence scientific judgment.

It is against this backdrop of challenge and opportunity that the publisher offers this supplementary text, *Of Pandas and People: The Central Question of Biological Origins.* The purpose of this text is to expose your students to the captivating and the controversial in the origins debate—to take them beyond the pat scenarios offered in most basal texts and encourage them to grapple with ideas in a scientific manner.

Pandas does this in two ways. First, it offers a clear, cogent discussion of the latest data relevant to biological origins. In the process, it rectifies many serious errors found in several basal biology texts.

Second, *Pandas* offers a different interpretation of current biological evidence. As opposed to most textbooks, which present the more-or-less orthodox neo-Darwinian accounts of how life originated and diversified, *Pandas* also presents a clear alternative, which the authors call "intelligent design." Throughout, the text evaluates how well different views can accommodate anomalous data within their respective interpretive frameworks. *Pandas* also makes the task of organizing your lessons and researching the scientific issues much easier. *Pandas* provides the scientific information you need and provides it in a way that coordinates well with your basal text.

In the spirit of good, honest science, *Pandas* makes no bones about being a text with a point of view. Because it was intended to be a supplemental text, the authors saw no value in simply rehashing the orthodox accounts covered by basal textbooks. Rather, its presentation of a non-Darwinian perspective, in addition to the standard view, is intended to stimulate discussion and encourage students to evaluate the explanatory power of different theories—which, after all, is what science is all about.

By using this text in conjunction with your standard basal text, you will help your students learn to grapple with multiple competing hypotheses and to maintain an open but critical posture toward scientific knowledge. As students learn to weigh and sort competing views and become active participants in the clash of ideas, you may be surprised at the level of motivation and achievement displayed by your students.

The Status of Evolution as a Fact

Despite the great value of presenting opposing viewpoints, the popular debate over origins has fostered several misconceptions about evolution, design, and science itself. To get the most benefit from this supplement, teachers should understand these misconceptions and be prepared to face them in an open and fair-minded manner when they arise.

One misconception concerns the status of evolution as a fact. In the origins debate, it is common to hear the assertion that evolution is not merely a theory but an indisputable fact.[8] Educators who take this view argue that it is futile and misleading to present non-Darwinian views as serious alternatives to Darwinian evolution.

The factual status of evolution, however, depends critically on what the word "evolution" means. Yale biologist Keith Stewart Thomson points out that scientists have used the term in at least three different ways.[9]

The first meaning he identifies is "change over time." In this sense, to say that evolution has taken place is to say that change has occurred and that things are different now from what they were in the past. The fossil evidence, for example, reveals different organisms from one geological period to the next.

When the word is used in this sense, it is hard to disagree that "evolution" is a fact. The authors of this volume certainly have no dispute with that notion. *Pandas* clearly teaches that life has a history and that the kinds of organisms present on earth have changed over time.

The second meaning that Thomson identifies is descent with modification—the idea that all organisms are "related by common ancestry."[10] Evolution in this sense is a theory about the history of life. In Darwin's view, that history can best be depicted as a single branching tree—a genealogical tree—in which life diversifies over time.

Many people assert that evolution in this second sense is a fact, just as gravity is a fact. But the two situations are hardly analogous. The fact of gravity can be verified simply by dropping a pencil—an experiment anyone can perform. Common ancestry, however, cannot be directly verified by such an experiment. We can no more "see" evolution in the fossil record than paleontologists of Darwin's day could "see" creation events. The best we can do is infer what might have happened in the past by piecing together circumstantial evidence from many different fields.

Darwin, for example, sought to establish common descent by examining evidence from several different areas: paleontology, biogeography, comparative anatomy, and embryology. Others have relied, in addition, on evidence from genetics, molecular biology, and biochemistry.

The problem with this kind of historical detective work, however, is that it seldom produces a conclusion that forecloses other alternatives. As philosopher of biology Elliot Sober points out, there may be any number of plausible explanations—or "past histories"—that can account for the same evidence.[11] Sober's observation recalls the insightful warning of fictional detective Sherlock Holmes. "Circumstantial evidence is a very tricky thing," said Holmes. "It may seem to point very straight to one thing, but if you shift your point of view a little, you may find it pointing in an equally uncompromising manner to something different."[12]

To use another analogy, consider the picture below.

What do you see—a tired old woman or an attractive young lady? Depending on what you're looking for, you can probably see either one. The drawing hasn't changed. The only thing that's different is the way you look at it.

The point is, unless we can eliminate plausible competing explanations, it's presumptuous to call descent with modification a fact. As most people understand the term, a fact "is supposed to be distinguished from transient theories as something definite, permanent, and independent of any subjective interpretation by the scientist."[13] By this definition, descent with modification simply doesn't warrant the status of a fact. Far from compelling a single conclusion, the evidence may legitimately be interpreted in different ways, leading to several possible conclusions. None of those conclusions warrants the status of a "fact".

As zoologist Thomas Kemp warns:

All attempts to understand the diversity of

> organisms rely upon empirically untestable assumptions either about evolution or about natural patterns. There is nothing wrong with making assumptions or seeking to justify them of course. It is the very stuff of science. What is unforgivable is to forget that they are assumptions and behave as if they were known certainties when they are no such things.[14]

Indeed, calling common descent a fact only closes off debate and blurs the distinction between fact and inference. That, in turn, makes us particularly vulnerable to the illusion that we know more than we really do. In the preface to his best-selling volume, *The Discoverers*, historian Daniel Boorstin tells the reader:

> The obstacles to discovery—the illusions of knowledge—are also part of our story. Only against the forgotten backdrop of the received common sense and myths of their time can we begin to sense the courage, the rashness, the heroic and imaginative thrusts of the great discoverers. They had to battle against the current "facts" and dogmas of the learned.[15]

This is precisely why a book that questions the Darwinian notion of common descent is so necessary. By presenting a reasonable alternative to evolution in this second sense (i.e., common ancestry), *Pandas* helps students learn to work with multiple perspectives, to distinguish those perspectives from facts, and to guard themselves against the illusion of knowledge.

The final meaning of evolution that Thomson identifies concerns the mechanism of biological change—the particular explanation of how evolution in the first two senses occurred. Here the term "evolution" refers to random variation and natural selection. In Thomson's words:

> Although many biologists act as though (the mechanism) were the whole meaning of evolution, it obviously is not. The first and second meanings could be explained by several different theories, and both had a serious intellectual history before 1859, while the third meaning is currently confined to a particular explanatory hypothesis, Darwinism.[16]

Evolution in this third sense asserts that the cause or mechanism of biological change is purposeless, nonintelligent, and completely naturalistic.[17] Oxford zoologist Richard Dawkins defended this view in his best-selling book, *The Blind Watchmaker.*[18] Like Darwin himself, Dawkins acknowledges that biological organisms appear to exhibit remarkable design. Yet both men claim that this appearance is an illusion, produced entirely by random variation and natural selection. Blind nature mimics intelligent design.

This "blind watchmaker" thesis is often touted as a fact, but it is not. For one thing, Darwinists have never demonstrated empirically that natural processes can create the complex structures that characterize living organisms. Like common descent, the blind watchmaker thesis is based on indirect evidence. It accounts for hypothetical transformations by extrapolating small observed changes over immense periods of time. Thus, the blind watchmaker thesis is not a fact, but an inference.[19]

What's more, the blind watchmaker thesis—at least in its neo-Darwinian form—may not be a warranted inference. As we mentioned at the beginning of this essay, neo-Darwinism has come under increasing attack from scientists and philosophers alike. Scientists have increasingly questioned the ability of mutation and natural selection to generate new organs, limbs, or body plans?[20] A host of other problems have led biologists Mae-Wan Ho and Peter Saunders to say:

> Until only a few years ago, the 'synthetic' or 'neo-Darwinist' theory of evolution stood virtually unchallenged as the basis of our understanding of the organic world . . . . Today, however, the picture is entirely different. More and more workers are showing signs of dissatisfaction with the synthetic theory. Some are attacking its philosophical foundations . . . . Others have

> deliberately set out to work in just those areas in which neo-Darwinism is least comfortable, like the problem of gaps in the fossil record or the mechanisms of non-Mendelian inheritance. . . . Perhaps most significantly of all, there is now appearing a stream of articles and books defending the synthetic theory. It is not so long ago that hardly anyone thought this was necessary.[21]

Pandas gives students a much-needed opportunity to explore the evidence and arguments that have caused some scientists to doubt contemporary Darwinism. It examines evidence from such fields as biochemistry, genetics and paleontology—evidence that casts doubt on the sufficiency of purposeless processes to explain the appearance of new biological forms.

Going a step further, *Pandas* helps students understand the positive case for intelligent design. Following a growing number of scientists and philosophers, the authors argue that life not only *appears* to have been intelligently designed but that it actually was. Drawing on recent developments in molecular biology, the authors show that even simple organisms bear all the earmarks of designed systems.

The authors also discuss what scientists have learned by applying mathematics and information science to biology. These disciplines suggest the possibility of distinguishing natural systems from intelligently designed ones—and have led some scientists to conclude that the "coded genetic information" (or sequence specificity) of DNA, proteins, and the like, reflects the activity of a pre-existent intelligence.[22] While that conclusion is still controversial, a growing minority of scientists see it as a plausible alternative to the blind watchmaker thesis.[23]

By presenting the case for intelligent design the authors demonstrate that there are alternatives to the blind watchmaker thesis—and that evolution as a purposeless process is neither an indisputable fact nor the only inference supported by biological data.

In sum, then, only in the most trivial sense—change over time—can evolution be considered a fact. Far from being a legitimate reason for avoiding alternative views, the alleged "fact of evolution" underscores precisely why a book like *Pandas* is so necessary. If students are to achieve true scientific literacy, they must learn to distinguish fact from supposition. A curriculum that blurs this distinction serves neither the students nor society.

Science and the Laws of Nature

A second misconception revolves around the question of what makes a concept or explanation "scientific." In particular, some scientists and philosophers assert that the concept of intelligent design is inherently non-scientific. According to this view, science must explain things by using natural laws—not by invoking the special action of an intelligent agent. Thus, we no longer explain the orbit of a planet by saying that an angel pushes it through the heavens. We explain it with Newton's law of universal gravitation.

In the same way, design is ruled out-of-court because it invokes an intelligent agent rather than natural laws. Philosopher of science Michael Ruse, for example, has said:

> Science attempts to understand this empirical world. What is the basis for this understanding? Surveying science and the history of science today, one thing stands out: Science involves the search for order. More specifically, science looks for unbroken, blind, natural regularities *(laws)*. Things in the world do not happen in just any old way. They follow set paths, and science tries to capture this fact.[24]

There are serious problems with this view, however. One problem is that it ignores areas of scientific investigation where intelligent design is a necessary explanatory concept. The search for extraterrestrial intelligence (SETI) is one example. At the time of this writing, radio telescopes are scanning the heavens, looking for artificial radio signals that differ from the random signals generated by natural objects in space. If we were to limit science to the search for "unbroken, blind, natural regularities (laws)" we would have to say that SETI is unscientific—by definition.

Archaeology would meet the same fate. Archaeologists routinely distinguish manufactured objects (e.g., arrowheads, potsherds) from natural ones (e.g., stones), even when the differences between them are very subtle. These manufactured objects then become important clues in reconstructing past ways of life. But if we arbitrarily assert that science explains solely by reference to natural laws, if archaeologists are prohibited from invoking an intelligent manufacturer, the whole archaeological enterprise comes to a grinding halt.

A second problem with limiting science to blind, natural regularities is that it confuses *laws* with *explanations*—an error that philosopher of science William Alston calls "a 'category mistake' of the most flagrant sort."[25] Laws and explanations are often two different things.

Scientific explanations often invoke not only laws but causal events and actions. For example, consider the field of modern cosmology. Most cosmologists explain the features of our universe not only by reference to the laws of physics, but by reference to a single event: the Big Bang. The Big Bang explains why galaxies throughout the universe seem to be receding from each other. It also explains the presence of low-level radiation that seems to permeate space.[26] These phenomena cannot be explained solely by reference to physical laws or natural regularities. Rather, the critical explanatory feature (the Big Bang) is a one-time event that established the conditions responsible for the phenomena that we now witness.

Moreover, sometimes scientific laws are hardly relevant to our explanations at all—such as when we try to explain why things turned out one way rather than another. For instance, Newton's law of universal gravitation may tell us why the earth has a Newtonian orbit rather than a non-Newtonian one. But it doesn't explain why the earth follows its present orbit, instead of some other orbit that is equally compatible with Newton's law. That kind of explanation requires something else—namely, information about how the earth attained its present position and velocity.[27]

A similar example can be drawn from the field of historical geology. If a historical geologist wanted to explain the unusual height of the Himalayas, invoking natural laws would be of little use. Natural laws alone cannot tell us why the Himalayas are higher than, say, the Rocky Mountains. That would require discovering antecedent factors that were present in building the Himalayas but not in other mountain-building episodes.

Thus, scientific explanation not only involves laws but may also involve past causal events. If scientists could never invoke past events and causes, they could not explain many important phenomena.

Why is this important? Because ignoring the role of causal events in scientific explanation has created a false dichotomy between agency—or intelligent design—and the laws of nature. The fact that scientific explanations may invoke laws doesn't mean that agency is somehow ruled out. Rather, intelligent agents can alter causal events and introduce other contributing factors. Although intervention may alter the course of subsequent events—sometimes in novel and unexpected ways—it does not violate natural laws.

Indeed, the actions of intelligent agents are themselves causal events. Therefore, citing the action of agents may be necessary to explain many present phenomena. Imagine trying to explain Mt. Rushmore withour reference to sculptors. Law-like explanations involving only natural processes would completely miss the critical explanatory factor. That is why archaeologists, forensic scientists and historians often find it impossible to avoid postulating intelligent agency.

The notion that science explains solely by reference to natural law suffers from yet a third problem. In addition to confusing laws with explanations, it assumes a cookie-cutter view of science, in which all disciplines ask similar questions and use the same "scientific method." This belies the rich diversity of methods that scientists use to understand the natural world.

Several philosophers, for instance, have argued that a clear distinction exists between the "inductive sciences" and the "historical sciences".[28] These two broad categories ask different kinds of questions and use different kinds of methods. The inductive (or nomological) sciences, on the one hand, ask questions about how the natural world generally operates. Hence, a virologist may try to discover how a particular enzyme helps a virus infect its host. Or a crystalographer may try to determine the effects of weightlessness on crystal growth. In each case, scientists seek to uncover the regularities that characterize natural phenomena.

The historical sciences, on the other hand, ask different kinds of questions. Rather than trying to understand how the natural world operates, the historical sciences seek to understand how things came to be. One example, of course, would be the historical geologist who was seeking to explain the unusual elevation of the Himalayas. Another would be an evolutionary biologist seeking to explain the origin of giraffes. Still another would be the archaeologist seeking to reconstruct an ancient culture. Note that in each case the goal is not to find new laws or regularities but to reconstruct past conditions and events.

The importance of this distinction to our present discussion is that although postulating intelligent intervention is completely inappropriate in the inductive sciences, the same is not true in the historical sciences. In the inductive sciences, the whole point is to discover how the natural world normally operates on it's own, i.e., *in the absence of intelligent intervention.* Postulating an intelligent agent would thus contradict the implicit goal of research in the inductive sciences.

In the historical sciences, however, the goal is to reconstruct past events and conditions. Thus, there is no need to impose such restrictions. Quite the reverse. As we have seen, the explanation of certain artifacts or features may require reference to intelligence. Intelligent agents may have left traces of their activity in the natural world. The historical scientist need not turn a blind eye to them.

Hence, when investigating the origin of the living world, it may be perfectly acceptable—depending on the evidence—to hypothesize an intelligent designer.

Observability

A third misunderstanding concerns the scientific status of unobservable objects and events. Some philosophers and scientists claim that intelligent design is not scientific because it invokes an unobservable intelligent designer. To be scientific, they claim, a concept or idea must be testable. Because an intelligent designer is unobservable, theories of intelligent design are not testable—and hence not scientific.

It is by no means clear, however, that something is untestable—and hence unscientific—simply because it is unobservable. If this were the case, many accepted theories and concepts would have to be declared unscientific as well. Chemist J. C. Walton observes:

> The postulation of. . .external intervention (into nature by a designer) undoubtedly restores order, harmony and simplification to the data of physics and biology. (Yet) at present there is no unambiguous evidence. . .for the existence of the external entity, but this should not be regarded as a drawback. Many key scientific postulates such as atomic theory, kinetic theory of the applicability of wave functions to describing molecular properties were, and still are, equally conjectural. Their acceptance depended, and still depends, on the comparison or their predictions with observables.[29]

Also falling in this category are almost all theories in the historical sciences—theories that postulate conditions and events that occurred in the unobservable past. The Big Bang is one such theory.

Another, ironically, is neo-Darwinism. Although neo-Darwinism explains many observable features of the living world, it postulates unobservable objects and events. For example, the mutational events that allegedly produced reptiles, birds, mammals, and even humans have never been observed—nor will they ever be observed. Similarly, the transitional life forms that occupy the branching-points on Darwin's tree of life are also unobservable. Transitional forms exist now only as theoretical entities that make possible a coherent Darwinian account of how present-day species originated.

The unobservable character of Darwinism becomes especially plain when proponents try to reconcile the fossil evidence with their theory. As paleontologists now admit, the fossil evidence looks a good deal less "Darwinian" than they had previously acknowledged. [30] Indeed, as Harvard paleontologist Stephen Gould points out, the two outstanding features of the fossil record are "sudden appearance" and "stasis." At any given location, species tend to appear "suddenly," fully formed, and exhibit no directional change during their stay on earth. [31]

The standard neo-Darwinian explanation for these features is the imperfection of the fossil record; because fossilization occurs only under special circumstances, fossils give us only a rough sketch of evolutionary history. More recently, some have proposed that evolutionary change occurs rapidly and in small, isolated populations of organisms. Both explanations, however, invoke unobserved circumstances to explain unobserved fossil organisms. How can one observe a non-fossilization event that happened 100 million years ago?

Darwin himself realized that much of the evidence for his theory was indirect. Indeed, he spent long hours defending his practice of inferring the unobservable from the observable.

> I am actually weary of telling people that I do not pretend to adduce direct evidence of one species changing into another, but that I believe that this view in the main is correct because so many phenomena can be thus grouped and explained.[32]

If we accepted the principle that unobservable entities are inadmissible in science, we would have to reject not only Darwin's theory but his entire approach to scientific investigation. [33]

To be fair, some opponents of intelligent design would argue that the real problem is not unobservability but flexibility. The concept of an intelligent designer is simply too much of a "wild card"; it can explain anything. Put another way, the concept of an intelligent designer cannot be falsified.

Intelligent design is not unique in its flexibility, however. We have already seen how Darwinists handle the problem of the fossil record; they account for unobserved fossil forms by invoking unobserved geological processes. Indeed, the history of science shows that scientists have often offered *ad hoc* explanations to save a cherished theory. This problem is particularly pronounced in the historical sciences, where investigators must draw conclusions from incomplete or sketchy evidence.

Nevertheless, intelligent design is not so flexible that it cannot be falsified. The concept of intelligent design entails a strong prediction that is readily falsifiable. [34] In particular, the concept of intelligent design predicts that complex information, such as that encoded in a functioning genome, never arises from purely chemical or physical antecedents. Experience will show that only intelligent agency gives rise to functional information. All that is necessary to falsify the hypothesis of intelligent design is to show confirmed instances of purely physical or chemical antecedents producing such information.

Religion and Intelligent Design

A final misconception you may encounter is that intelligent design is simply sectarian religion. According to this view, intelligent design is merely fundamentalism with a new twist; teaching it in public schools allegedly violates the separation of church and state.

This view is wide of the mark. The idea that life had an intelligent source is hardly unique to Christian fundamentalism. Advocates of design have included not only Christians and other religious theists, but pantheists, Greek and Enlightenment philosophers and now include many modern scientists who describe themselves as religiously agnostic. [35] Moreover, the concept of design implies absolutely nothing about beliefs normally associated with Christian fundamentalism, such as a young earth, a global flood, or even the existence of the Christian God. All it implies is that life had an intelligent source.

In any case, sectarianism is more a matter of form than content. [36] It is marked by a certain narrowness and exclusivity that entertains no debate and tolerates no opposing viewpoints. Given the broad appeal of intelligent design (even Richard Dawkins, a staunch Darwinist and author of *The Blind Watchmaker*, acknowledges "the appearance of design" in the living world [37]), it is perhaps more accurate to conclude that the real sectarians are those who villify design as "fundamentalist religion." Such name-calling is merely another way to avoid debate and keep the real issues out of view. [38]

Even if the design hypothesis were religious, however, criticizing it on that basis begs the question of whether it is scientifically warranted. In science, the origin of an idea is supposed to be irrelevant to its validity. What matters is not the source but whether the idea is logically consistent and empirically supportable. If it is, what justification is there for excluding it from the classroom?

In its landmark ruling on the Louisiana Balanced Treatment Act, the United States Supreme Court did not try to shield the classroom from dissenting viewpoints. Indeed, it affirmed that teachers already had the flexibility to teach nonevolutionary views and present scientific evidence bearing on the question of origins:

> The Act does not grant teachers a flexibility that they did not already possess to supplant the present science curriculum with the presentation of theories besides evolution, about the origin of life. Indeed, the Court of Appeals found that no law prohibited Louisiana public school teachers from teaching any scientific theory. [39]

Neither did the Supreme Court choose to limit that flexibility:

> Teaching a variety of scientific theories about the origins of humankind to schoolchildren might be validly done with the clear secular intent of enhancing the effectiveness of science instruction. [40]

This is not only consistent with good science, it is consistent with the highest ideals of a democratic society. As John Scopes, who was tried in the 1920s for teaching evolution, said at his own trial, "Education, you know, means broadening, advancing, and if you limit a teacher to only one side of anything the whole country will eventually have only one thought, be one individual. I believe in teaching every aspect of every problem or theory." [41]

References

1. R.H. Brady, 1982. *Biological Journal of the Linnean Society* 17,79-96; D. Collingridge and M. Earthy, 1990. *History and Philosophy of the Life Sciences* 12,3-26; G. de Beer, 1971. *Homology: An Unsolved Problem. London: Oxford University Press; M. Denton, 1986 Evolution: A Theory in Crisis.* Bethesda, MD: Adler and Adler; P.P. Grasse, 1977. *Evolution of Living Organisms*. New York: Academic Press; M-W. Ho, 1965. *Methodological Issues in Evolutionary Theory.* Doctoral dissertation, Oxford University; F. Hoyle and S. Wickramasinghe, 1981. *Evolution From Space.* London: J.M. Dent; P.E. Johnson, 1991. *Darwin on Trial*. Washington, DC: Regnery Gateway; S. Lovtrup, 1987. *Darwinism: The Refutation of Myth*. Beckingham, Kent: Croom Helm Ltd.; D. Raup, 1979. *Field Museum of Natural History Bulletin*. 50 (1); R. Lewin, 1980. *Science* 210,883; R. Lewin, 1988. *Science* 241,291; R. Lewin, 1987. *Bones of Contention.* New York: Simon and Schuster; C. Mann, 1991. *Science* 252,378-381, esp. 379; P.S. Moorhead and M.M. Kaplan, 1967. *Mathematical Challenges to the neo-Darwinian Interpretation of Evolution*. Philadelphia: Wistar Institute Press. (See especially papers and comments from M. Eden, M. Shutzenberger, S.M. Ulam and P. Gavaudan); P.T. Saunders and M-W. Ho, 1982. *Nature and System* 4, 179-191; A. Tetry, 1966. *A General History of the Sciences* Vol. 4. London: Thames and Hudson. (see section on evolution, esp. p. 446.); G. Webster, 1984. *Beyond Neo-Darwinism*. London: Academic Press; and S.J. Gould, 1980. *Paleobiology* 6,119-130.
2. N. Eldredge, 1985. *Time Frames: The Evolution of Punctuated Equilibria*. Princeton: Princeton University Press, p. 14.
3. S.J. Gould, 1980. *Paleobiology* 6, 119-130.
4. W. Bradley, 1988. *Perspectives*, 40,72-83; K. Dose, 1988. *Interdiscipl. Sci. Rev.* 13,348-356; R.A. Kok, J.A. Taylor, and W.L. Bradley, 1988. *Origins of Life and Evolution of the Biosphere* 18,135-42; P.T. Mora, 1963. *Nature* 199,212-19; P.T. Mora, 1965. In S.W. Fox, ed. *The Origins of Prebiological Systems and of Their Molecular Matrices*. New York: Academic Press, pp. 39-64, 310-15; J.N. Moore, 1978. *J. Amer. Scient. Affil.* Special Edition on Evolution; H.J. Morowitz, 1966. In Wostenholme, O'Connor, and Churchill, eds. *Principles of Biomolecular Organisation,* London:Wiley, pp. 446-59; H.J. Morowitz, 1968. *Energy Flow in Biology*. New York: Academic Press; H.H. Pattee, 1970. In C.H. Waddington, ed. *Towards a Theoretical Biology.* Edinburgh: University Press 3,117-36; R. Shapiro, 1986. *Origins: a Skeptic's Guide to the Creation of Life on Earth*. New York: Summit Books; J.M. Smith, 1979. *Nature* 280,445-46; L. Margulis, J.C. Walker, and M. Rambler, 1976. *Nature* 264,620-24; S. Miller and J. Bada, 1988. *Nature* 334,609-10; C. Thaxton, W. Bradley, and R. Olsen, 1984. *The Mystery of Life's Origin*. Dallas: Lewis and Stanley; J.C. Walton, 1977. *Origins* 4,16-35; E. Wigner, 1961. In *The Logic of Personal Knowledge*, Essays presented to Michael Polanyi. London: Routledge and Kegan Paul, pp. 231ff; H.P. Yockey, 1977. *J. Theoret. Biol.* 67,377-98; and H.P. Yockey, 1981. *J. Theoret. Biol.* 91,13-31.
5. D. Collingridge and M. Earthy, 1990. *History and Philosophy of the Life Sciences* 12, 3-26; K. Dose, 1988. 13,348; See also quotation from Carl Woese in R. Shapiro, 1986 p. 114.
6. Ibid.
7. S.J. Gould, 1980.
8. See for example the 1990 California Science Framework, published by the California Department of Education. See also M. Ruse, 1982. *Darwinism Defended: A Guide to the Evolution Controversy*. London: Addison-Wesley, p. 58.
9. K.S. Thomson. *American Scientist*, Sept./Oct. 1982, pp. 529-531.
10. Ibid.
11. E. Sober, 1988. *Reconstructing the Past.* Cambridge, Mass: MIT Press, pp. 4-5. See also D.T. Campbell and J.C. Stanley, 1963. *Experimental and Quasi-Experimental Designs for Research*. New York: Houghton Mifflin.
12. Sir A.C. Doyle, *The Boscome Valley Mystery* quoted in G.P. Capretti, 1983; Peirce, Homes, and Popper in U. Eco and T. Sebeok, eds. *The Sign of Three*. Bloomington: Indiana University Press, p. 145.
13. L. Fleck, 1979. *Genesis and Development of a Scientific Fact.* Trans. by Thomas Merton. Chicago: University of Chicago Press, p. xxvii. Actually, Fleck argues that there is really no such thing as a "fact" in this sense. All "facts" involve a certain amount of subjective interpretation. If this is so, it strengthens the case against the "fact" of evolution (in the sense of common descent). Nevertheless, we still believe that the fact/inference distinction is a useful one, underscoring as it does the difference between ideas in which we have great confidence, and those that seem less sure.
14. T. Kemp, 1985. In *Oxford Surveys of Evolutionary Biology*, Vol. 2, R. Dawkins and M. Ridley, eds. New York: Oxford University Press, p. 153.
15. D.J. Boorstin, 1985. *The Discoverers*. New York: Vintage Books, p. xv.
16. Thomson, 1982, p. 531.
17. Ibid., pp. 530-531.
18. R. Dawkins, 1986. *The Blind Watchmaker*. New York: W.W. Norton.
19. See for example Ridley, 1985. pp. 3-8.
20. J. Webster and B. Goodwin, 1982. *Journal of Social and Biological Structures*, 5,15-47; D. B. Wake and G. Roth, eds. 1989. *Complex Organismal Functions*. New York: John Wiley; K. Padian, 1989. *Paleobiology* 1,73-78; R.A. Raff and E.C. Raff, eds. 1987. *Development as an Evolutionary Process*. New York: Alan R. Liss, p. 84; S. Kaufman, 1985. *Cladistics* 1,247-265; K.S. Thomson, 1988. *Morpho Genesis and Evolution*. New York: Oxford University Press; M-W. Ho and P.T. Saunders, 1979. *J. Theoret. Biol.* 78,573-591; B. John and G. Miklos, 1988. *The Eukaryote Genome in Development*. London: Allen & Unwin.
21. M-W. Ho and P.T. Saunders, eds. 1984. *Beyond Neo-Darwinism*. London: Academic Press, p. ix.
22. See especially C. Thaxton, W. Bradley, and R. Olsen, 1984; E.J. Ambrose, 1982. *The Nature and Origin of the Biological World*. New York: Wiley, Halsted; Denton, 1986; Walton, 1977.
23. See, for example, W.H. Thorpe, 1978. *Purpose in a World of Chance: A Biologists's View.* New York: Oxford University Press; F. Hoyle, 1983. *The Intelligent Universe*. New York: Holt, Reinhart & Winston; D.A. Kuznetsov, 1989. *International Journal of Neuroscience* 49,43-59; H. Yockey, 1992. *The Mathematical Foundations of Molecular Biology*. New York: Cambridge University Press; C. Thaxton, W. Bradley, and R. Olsen, 1984. *The Mystery of Life's Origin*. Dallas: Lewis and Stanley; E. J. Ambrose, 1982. *The Nature and Origin of the Biological World*. New York: Wiley, Halsted;' R. Bohlin and L. Lester, 1984. *The Natural Limits to Biological Change*. Grand Rapids: Zondervan.
24. M. Ruse, 1982. *Science, Technology, and Human Values*, 7,72-73.
25. W.P. Alston, 1971. *Philosopohy of Science* 38,13-34.
26. For a readable and engaging discussion of the Big Bang, and the controversy this concept engendered, see R. Jastrow, 1978. *God and the Astronomers*. New York: W.W. Norton.
27. P. Lipton, 1991. *Inference to the Best Explanation*. London: Routledge. p. 52.

28. Philosophers have adopted many different terms to distinguish between these two kinds of science. We have adopted the terms "inductive science" and "historical science" because they are less cumbersome than some of the other terms we could have used. For a fuller discussion of the distinction between these two classes of science, see S. Meyer, 1990. *Of Clues and Causes: a Methodological Interpretation of Origin of Life Studies*. Doctoral dissertation, University of Cambridge.
29. J.C. Walton, 1977.
30. "So, here's a bit of a dilemma. When we finally find some evolutionary change, however slight it may seem, the "typostrophic" sort of affair the *Phacops rana* lineage seems to show in the Midwest poses a choice between two unappetizing alternatives: either you stick to conventional theory despite the rather poor fit of the fossils, or you focus on the empirics and say that saltation looks like a reasonable model of the evolutionary process—in which case you must embrace a set of rather dubious biological propositions. Paleontologists are rather well known for taking that latter course—adopting *ad hoc*, outmoded and sometimes downright mystical ideas about biological processes just because they fancy these ideas fit what they think they see in the fossil record. I had every desire to avoid that well-trodden path. Besides, I was (and remain) too much of a conventional neo-Darwinian ever to subscribe to the saltationist heresy." From Eldredge, 1985. p. 75.
31. S. J. Gould, 1979. *Natural History* 86(5).
32. F. Darwin, 1903. *More Letters of Charles Darwin*, Vol. 1. New York: Appleton, p. 184.
33. This is exactly what Darwin's critics did. Many of them rejected his theory precisely because it was based on "non-scientific" reasoning. The Darwinian revolution was at least partly a revolution in what people considered to be "scientific." See N. Gillespie, 1979. *Charles Darwin and the Problem of Creation*. Chicago: University of Chicago Press; and D. Hull, 1973. *Darwin and His Critics*. Cambridge, Mass: Harvard University Press.
34. This kind of prediction is called a *proscriptive generalization*. Proscriptive generalizations make strong statements about what will *not* happen if a scientific theory is true. They describe phenomena that the theory proscribes. Thus, the laws of thermodynamics predict that we will never witness any instance of perpetual motion. That's a proscriptive generalization. Because proscriptive generalizations make such strong statements, they are readily falsifiable.
35. M. Denton, 1986. *Evolution: A Theory in Crisis*. Bethesda, MD: Adler and Adler; J.D. Barrow and F.J. Tippler, 1986. *The Anthropic Cosmological Principle*. Oxford. Clarendon.
36. J.D. Hunter, 1992. *Origins Reserach* 14,1-14. See also J.D. Hunter, 1991. *Culture Wars: The Struggle to Define America*. New York: Basic Books.
37. "Natural selection is the blind watchmaker, blind because it does not see ahead, does not plan consequences, has no purpose in view. Yet the living results of natural selection overwhelmingly impress us with the appearance of design as if by a master watchmaker, impress us with the illusion of design and planning. The purpose of this book is to resolve this paradox to the satisfaction of the reader, and the purpose of this chapter is to further impress the reader with the power of the illusion of design. We shall look at a particular example and shall conclude that, when it comes to complexity and beauty of design, Paley hardly even began to state the case." From R. Dawkins, 1986. *The Blind Watchmaker*. New York: W.W. Norton.
38. For a good discussion of this exclusivism, see P.E. Johnson, 1991. *Darwin on Trial*. Washington, DC: Regnery Gateway.
39. *Edwards v. Aguillard*, 482 U.S. (June 19, 1987).
40. *Edwards v. Aguillard*, p. 14.
41. Cited in P. Davis, and E. Solomon, 1973. *The World of Biology*. New York: McGraw Hill, p. 610.

INDEX

Authors and Contributors

Percival Davis—Coauthor. Professor of Life Science, Hillsborough Community College, Tampa, Florida, since 1968; author of several college level biology texts, including *Biology* with Claude Villee and Eldra Solomon (W. B. Saunders, 1985); B.A. in zoology from DePauw University; M.A. in zoology from Columbia University; 60 credit hours beyond the Master's degree at Columbia University and the University of South Florida in zoology, ecology and physiology.

Dean H. Kenyon—Coauthor. Professor of Biology, San Francisco State University, San Francisco, California; contributing author to festschrifts of A. I. Oparin and Sidney Fox; coauthor of *Biochemical Predestination* (McGraw-Hill, 1969), which was the best-selling advanced level book on chemical evolution in the 1970s; S.B. in physics, 1961 from the University of Chicago; Ph.D. in biophysics, 1965 from Stanford University; National Science Foundation Postdoctoral Fellow 1965–1966 at the University of California, Berkeley; visiting scholar in 1974 to Trinity College, Oxford University; Associate, Chemical Evolution Branch, NASA-Ames Research Center in California, 1974–1976; Phi Beta Kappa.

Charles B. Thaxton—Academic Editor. B.S. in chemistry, 1962 from Texas Tech University; M.S. in chemistry, 1964 from Texas Tech University; Ph.D. in physical chemistry, 1970 from Iowa State University; post doctoral appointments at Brandeis University in Molecular Biology and at Harvard University in the History and Philosophy of Science; Fellow of the American Institute of Chemists; co-author of the best-selling college text on chemical evolution, *The Mystery of Life's Origin* (Lewis and Stanley, Dallas, 1984).

Mark D. Hartwig—Teacher Introduction. Ph.D. in educational psychology from the University of California at Santa Barbara, with an emphasis in statistics and research methodology; employed by UCSD for seven years as a research and evaluation specialist; since 1989, Executive Director of Access Research Network, which serves as a public source of information on science, technology, and society; news editor for *Science Probe!*, a national newsstand magazine.

Stephen C. Meyer—Teacher Introduction. Ph.D. from Cambridge University in the history and philosophy of science, thesis on the logical structure of evolutionary arguments and the methodology of the historical sciences; B.S. in geology and physics; recipient of a Rotary International Scholarship and three-year fellowship from the American Friends of Cambridge University; four years of experience as a geophysicist with the Atlantic Richfield Company; author of editorial features on scientific issues for Los Angeles Times and Wall Street Journal; teaches History and Philosophy of Science at Whitworth College in Spokane, Washington.

CREDITS

(Page 2) From the book *The One Hundred* published by arrangement with Carol Publishing Group

(Page 3) Reproduced with permission from the University of California, San Diego

(Page 12) Supplied by Carolina Biological Supply Company

(Page 12) Photograph from p. 26 of *Giraffe* by Caroline Arnold, photographs by Richard Hewett. Text copyright ©1987 by Caroline Arnold, photographs copyright ©1987 by Richard Hewett. Reprinted by permission of Morrow Junior Books, a division of William Morrow & Co., Inc.

(Page 13) THE FAR SIDE cartoon by Gary Larson is reprinted by permission of Chronicle Features, San Francisco, CA

(Page 14) ©Daniel Grogan/Uniphoto

(Page 15) Photo of Hemignathus munroi courtesy of Jack Jeffrey

(Page 20) Photo by Jan Clark

(Page 23) Negative no. 326672, courtesy Department of Library Services, American Museum of Natural History

(Page 23) ©Tom McHugh, The National Audubon Society collection/PR

(Page 24) Photo by Dan Smith, courtesy of Vertebrate Paleontology Lab, Texas Memorial Museum, University of Texas at Austin

(Page 31) Photo courtesy of the Dallas Zoo

(Page 32) ©New York Zoological Society

(Page 33) Photo courtesy of Al Buell

(Page 35) Photo courtesy of Jon R. Buell

(Page 36) Photo courtesy of James A. Kern

(Page 52) From S.W. Fox and K. Dose, *Molecular Evolution and the Origin of Life*, revised ed., p. 214, Marcel Dekker, Inc. NY

(Page 83) Reproduced with permission from the *Annual Reviews of Ecology and Systematics*, Vol. 12 ©1981 by Annual Reviews, Inc.

(Page 83) Photo courtesy of James L. Nation, Jr.

(Page 101) From Alfred S. Romer, *Vertebrate Paleontology* ©1966. Reprinted by permission of The University of Chicago Press

(Page 102) Courtesy of the Natural Peabody Museum at Yale University

(Page 103) Negative no. 3211683. Courtesy Department of Library Services, American Museum of Natural History

(Page 105) From *The Origin of Birds* by Gerhard Heilmann ©1927 by D. Appleton and Co., renewed ©1955 by Gerhard Heilmann. Used by permission of the publisher, Dutton, an imprint of the New American Library, a division of Penguin Books USA Inc.

(Page 119) ©New York Zoological Society

(Page 119) Photo courtesy of the Dallas Zoo

(Page 120) Courtesy Department of Library Services, American Museum of Natural History

(Page 123) ©S. J. Krasemann—Peter Arnold, Inc.

(Page 125) ©Tom McHugh, The National Audubon Society Collection/PR

(Page 126) Photo courtesy of NASA Ames Research Center

(Page 131) Supplied by Carolina Biological Supply Company

(Page 132) Figure from *Principles of Anatomy and Physiology*, 5/ed., by Gerard Tortora and Nicholas P. Anagnostakos. Copyright ©1987 by Biological Sciences Textbooks, Inc., A and P Textbooks, Inc., and Elia-Sparta, Inc. Reprinted by permission of Harper Collins Publishers

(Page 140) Reprinted with permission from W. H. Freeman and Company